The Colours of Infinity

The Beauty and Power of Fractals

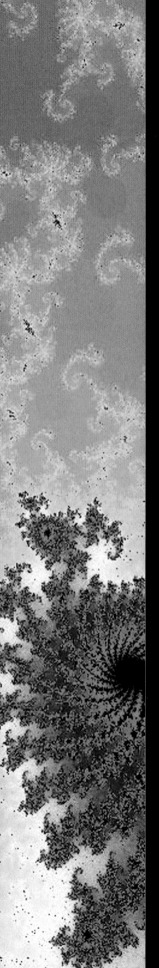

The Colours of Infinity

The Beauty and Power of Fractals

With contributions by

Ian Stewart, Sir Arthur C. Clarke,

Benoît Mandelbrot, Michael and Louisa Barnsley,

Will Rood, Gary Flake and David Pennock

and Nigel Lesmoir-Gordon

 Springer

Editor
Nigel Lesmoir-Gordon
Gordon Films UK
MK45 4AD Bedfordshire
United Kingdom
nigel@gordonfilms.tv

Additional material to this book can be downloaded from http://extras.springer.com
Password: [978-1-84996-485-2]
ISBN 978-1-84996-485-2   e-ISBN 978-1-84996-486-9
DOI 10.1007/978-1-84996-486-9
Springer London Dordrecht Heidelberg New York

British Library Cataloguing in Publication Data
A catalogue record for this book is available from the British Library

Library of Congress Control Number: 2010937321
Mathematics Classification Number (2010) 28A80

Printed on acid-free paper

Springer is part of Springer Science+Business Media (www.springer.com)

This book is dedicated
to Benoît Mandelbrot

'The most beautiful thing
we can experience is the mysterious.
It is the source of all true
art and science.'

Albert Einstein

Contents

Introduction

A geometry able to include mountains and clouds now exists. I put it together in 1975, but of course it incorporates numerous pieces that have been around for a very long time. Like everything in science, this new geometry has very, very deep and long roots.

Benoît B. Mandelbrot

Introduction

This enhanced and expanded edition of THE COLOURS OF INFINITY features an additional chapter on the money markets by the fractal master himself, Professor Benoît Mandelbrot. The DVD of the film associated with this book has been re-mastered especially for this edition with exquisite new fractal animations, which will take your breath away!

Driven by the curious enthusiasm that engulfs many fractalistas, in 1994, Nigel Lesmoir-Gordon overcame enormous obstacles to raise the finance for, then shoot and edit the groundbreaking TV documentary from which this book takes its name. The film has been transmitted on TV channels in over fifty countries around the world. This book is not just a celebration of the discovery of the Mandelbrot set, it also brings fractal geometry up to date with a gathering of the thoughts and enthusiasms of the foremost writers and researchers in the field.

As Ian Stewart makes clear in the opening chapter, there were antecedents for fractal geometry before 1975 when Mandelbrot gave the subject its name and began to develop the underlying theory. It took the genius of Mandelbrot, allied with the computer power available to him at IBM, to realize the practicality, beauty and fascination in the subject, and to act as its propagator through a long and influential career.

The first chapter by Benoît Mandelbrot in this book is based on a paper delivered before a Nobel Conference in Stockholm called A Geometry Able to Include Mountains and Clouds. The breadth of his vision, extending from mathematics to economics, from art to language, is extraordinary. As several of the contributors note, once you take a fractal view of the universe, you see the evidence everywhere – in water, in clouds, in trees, in art (see Rood's chapter), in the human body and in the workings of the World Wide Web (Flake and Pennock). Mandelbrot's second chapter, Fractal Financial Fluctuations looks deeply into the fractal nature of the growth and collapse of financial prices. His radically new fractal modelling techniques cast a whole new light of order into the seemingly impenetrable thicket of the financial markets.

The article by Arthur C. Clarke is a special case. Its 4,000 or so words are a lucid miniature of scientific popularization, reflecting the excitement fractal geometry induces in so many of its converts. It also, as Nigel

Lesmoir-Gordon explains in his account of how the film came to be made, offered a link between himself and Clarke, the film's anchor, and lent its name to the film project itself.

Four of the film's contributors (Stewart, Clarke, Mandelbrot and Barnsley) have chapters in the book. Rood, Flake and Pennock, as well as Nigel Lesmoir-Gordon, the film's begetter, contribute original chapters specifically for this volume.

Using a metaphor of a random soccer game, Michael Barnsley with his wife Louisa, the originators of fractal image compression technology, present the ideas of fractal transformation and colour stealing using random iteration for the first time.

Will Rood takes the animation of fractals into a new area by explaining how the M-set is coloured and then how the strange reptiles of Dutch conceptual artist M. C. Escher (1898–1972), the 'undisputed master of tessellated art', can be mapped onto the exterior of quadratic fractals, allowing the creation of tessellation with fractal limits.

Gary Flake and David Pennock propose an 'optimistic and realistic' interpretation of the NFL ('no free lunch') theory as a key to understanding the current state of the World Wide Web and how it will evolve over time. Given its huge traffic and lack of central authority, the Web could have been infinitely complex, but it is in fact exceedingly regular; and this regularity can be exploited to make more effective algorithms for finding information on the Web.

The Colours of Infinity brings together all the leading names in the fractal geometry field. Between them the contributors have published at least 200 books under their own names, and in collaboration. You will feel in their articles an ease with communicating sometimes difficult abstract concepts and an urge to share the powerful meanings their insights into the world of fractals have for all of us. In terms of positive energy and commitment to the subject they are a persuasive community.

The last chapter of this collection is unusual in that it sets out the full shooting script of the film, with audio and spoken word alongside. This may well prove invaluable source material in, for example, the educational use of the film, which has gradually increased over the decade or so since the film's release.

The Colours of Infinity, the movie, made with so much evident pleasure, is approaching cult status and now gains a new lease of life by being coupled with this stimulating collection, expanding the film's concerns still further.

The soundtrack of the DVD, with Pink Floyd's David Gilmour's soaring guitar almost an aural fractal in its own right, is totally accessible, as are Will Rood's beautifully coloured animations of the fractals. The music and the images together have become club and garage favourites, and it is easy to understand why. Is it too far fetched to see in this harmonious matching of sound and image a tribute to the way Stanley Kubrick handled them in the Stargate sequence of his science fiction masterpiece *2001: A Space Odyssey?* – a powerful link back to Arthur C. Clarke.

One of the many strange thoughts that the M-set generates is this. In principle, it could have been discovered as soon as the human race learned to count. In practice, since even a low magnification image may involve billions of calculations, there was no way in which it could even be glimpsed before computers were invented.

Sir Arthur C. Clarke

Contributors

Arthur C. Clarke
Professor Physical Research Laboratory, Ahmedabad, Gujarat India

Benoît Mandelbrot (Emeritus)
Stirling Professor Department of Mathematics, Yale University, New Haven, CT, USA

Dr. David Pennock
Research Scientist at Yahoo!, Pasadena, CA, USA

Dr. Gary Flake
President Yahoo! Research Labs, Sunnyvale, CA, USA

Ian Stewart (Emeritus)
Professor Department of Mathematics, Warwick University, Coventry, Warwickshire, UK

Michael Barnsley
Professor Mathematical Sciences Institute, Australian National University, Canberra, Australia

Nigel Lesmoir-Gordon
Director Gordon Films, Clophill, Bedford, UK

Will Rood
Freelance Software Designer and Filmmaker London, UK

1 The Nature of Fractal Geometry

Ian Stewart

Fractals are more than just stunning visual effects – they open up new ways to model nature and allow us to quantify terms like 'irregular', 'rough' and 'complicated', writes mathematician Ian Stewart. His chapter does a service to the non-specialist reader in giving a largely non-technical introduction to fractal geometry in the context of mathematical traditions and its commercial applications. Stewart shows both how concepts like fractal dimension have a lengthy prehistory and also how Mandelbrot brought to the subject a systematic treatment, uniting theory and application. Mandelbrot's most important contribution to fractal geometry, Stewart suggests, 'was the realization that there was a subject'.

Ian Stewart is Emeritus Professor of Mathematics at Warwick University. In 1995 he was awarded the Royal Society's Michael Faraday Medal for furthering the public understanding of science. He was elected a Fellow of the Royal Society in 2001, and won the Public Understanding of Science and Technology Award of the American Association for the Advancement of Science in 2002.

He is the author of over 60 books including *Nature's Numbers, Does God Play Dice?, Figments of Reality, The Magical Maze, Life's Other Secret, What Shape is a Snowflake?, Evolving the Alien*, the best-sellers *The Science of Discworld I and II* (with Terry Pratchett and Jack Cohen), and the US best-seller *Flatterland*. He has also written a science fiction novel, *Wheelers* (with Jack Cohen).

N. Lesmoir-Gordon (ed.), *The Colours of Infinity: The Beauty and Power of Fractals*,
DOI 10.1007/978-1-84996-486-9_1, © Springer-Verlag London Limited 2010

The universe is full of fractals. Indeed it may even be one.

Thirty years ago, no one had heard of fractals. The concept existed, but the name was not coined until about 1975. Today, almost everyone has heard of fractals, and probably has a mug or a T-shirt or a poster somewhere around the house with one of the remarkable, intricate computer images that the word brings to mind. The importance of fractals, however, goes well beyond their visual attractiveness. What makes them so useful in today's scientific research is that they have opened up entirely new ways to model nature. They give scientists a powerful tool with which to understand processes and structures hitherto described merely as 'irregular', 'intermittent', 'rough', or 'complicated'.

What is a fractal? As a first, broad-brush description: it is a geometric form that possesses detailed structure on a wide range of scales. Think of the rocky slopes of a mountain, the proliferating fronds of a fern, and the fluffy outline of a cloud. These are physical objects: 'fractal' is a mathematical concept, and it relates to the real world in the same manner that 'sphere' relates to the shape of the Earth and 'spiral' relates to the shape of a snail shell. A mathematical fractal idealizes the intricacy of rocks and clouds: it has detailed structure on all scales. However much it is magnified, it does not 'flatten out' into a simple shape like a line or a plane.

Mathematical objects are idealized models of certain features of the real world; they are not real things, and they do not correspond exactly to real things. The Earth is not a perfect sphere; even allowing for its bulging equator, it is not a perfect ellipsoid either, even though many astronomy and earth science textbooks describe it that way. It has mountain ranges that give it a rough surface, unlike the infinite smoothness of the mathematical ideal. However, this type of inaccuracy does not stop scientists modelling the Earth as a sphere. In fact, the great advantage of a sphere as a model, for many purposes, is that it does not represent the intricacies of the real planet exactly. If it did, it would be no more use than a map of New York that is the same size as New York, with every traffic-light, doorstep, and cat rendered in perfect detail. A map must be simpler than the territory.

Models are tailored to suit particular objectives. If the objective is to understand mountain-building, then it is pointless to assume that the Earth is a smooth sphere. But if the objective is the long-term behaviour of the solar system, then a sphere is entirely acceptable, and a 'point mass' – even further from physical reality, since it assumes the Earth's diameter is zero – may well be better. In the same way, a mathematical fractal has detailed structure on scales so fine that they subdivide atoms – indeed, on scales finer than the Planck Length, at which level the universe becomes lumpy instead of smooth and 'distance' makes no sense. This discrepancy with the real world does not make fractals useless or irrelevant. As with the sphere and the map, what matters is the extent to which the model illuminates reality, not the extent to which it copies reality.

Fractals make it possible to quantify terms like 'irregular', 'intermittent', 'rough', and 'complicated'. How rough? 1.59 rough or 2.71 rough? Fractal geometry gives such statements a meaning, and makes it possible to test them in experiments. Mathematics provides a number, associated with each fractal, called its fractal dimension. The dimension reflects, among other things, the scaling properties of the fractal – how its structure changes when it is magnified. Unlike the traditional smooth curves and surfaces of much mathematical physics and applied mathematics, the dimension of a fractal need not be a whole number. It can, for example, be 1.59 or 2.71.

In fact, the difference between the fractal dimension of a geometric shape and its dimension in the usual 'topological' sense of mathematics provides a quantitative measure of just how rough the fractal is.

The notion of a fractal was brought to scientific prominence by Benoît Mandelbrot in 1975, and promoted in his book Fractals: Form, Chance, and Dimension of 1977. A revised edition appeared in 1982 under the title The Fractal Geometry of Nature. The term 'fractal' was introduced by Mandelbrot, but many of the subject's concepts – notably fractal dimension – have a lengthy prehistory. Mandelbrot's contributions to the subject have been many, but the most important was the realization that there was a subject. Mathematicians had studied spaces of non-integer dimension long before Mandelbrot; scientists had observed scaling laws and self-similarities in natural phenomena. But a systematic treatment, uniting theory and application, was lacking.

Now, some thirty years later, the theory that was stimulated by Mandelbrot's insight is thriving. A glance through the leading scientific journals, such as Nature and Science, will make it clear that fractals have become a standard technique of scientific modelling in a wide variety of areas. The mere existence of fractal structures immediately suggests a wide range of physical and mathematical questions, by directing our attention away from the classical obsession with smooth curves and surfaces. What happens to light waves passing through a medium whose refractive index is fractally distributed? Reflected in a fractal mirror? What sounds will a drum make if it has a fractal boundary? Traditional methods have little to say about such questions.

The importance of fractals

Are they important? Undoubtedly. Turbulence in the atmosphere makes it difficult for Earth-based telescopes to produce accurate images of stars; a turbulent atmosphere is well modelled by a fractal distribution of the refractive index. Light bouncing off the ocean, with its myriad waves on many scales, closely resembles reflection from a fractal mirror. And the way trees absorb energy from the wind is closely related to the 'vibrational modes' of a fractal – and it is such modes that create the sound of a drum. The natural world provides an inexhaustible supply of important problems in fractal physics. Already, technological and commercial advances have stemmed from such questions – for example, a compact antenna for mobile phones, new ways to analyse the movements of the stock market, and efficient methods to compress the data in computer images, squeezing more pictures onto a CD.

Once our eyes have been opened to the fact that fractal objects possess a distinctive character and structure, and are not just irregular or random, it becomes obvious that the universe is full of fractals. Indeed, it may even be one. Fractals teach us not to confuse complexity with irregularity, and they open our eyes to new possibilities. Fractals represent an entire new regime of mathematical modelling, which science is just beginning to explore.

The term 'fractal' was introduced by Mandelbrot, but many of the subject's oncepts – notably fractal dimension – have a lengthy prehistory. Mandelbrot's contributions to the subject have been many, but the most important was the realization that there was a subject.

Gallery of monsters
The prehistory of fractals

The prehistory of fractals goes back over a hundred years, to when mathematicians began thinking about new kinds of curves and surfaces, totally different from the shapes typically studied in classical geometry. The classical shapes are lines and planes, cones and spheres, curves and surfaces – and, except for the occasional edge or corner, these curves and surfaces are smooth and very well behaved. Smoothness in effect implies that they have no interesting small-scale structure: when magnified sufficiently, they appear flat and featureless. This absence of structure on small scales is crucial to classical 'limiting' analysis – the time-honoured methods of the calculus, which go back to Isaac Newton and Gottfried Leibniz. The very methodology of the calculus, the central technique of physics for more than two centuries, is to approximate a curve by its tangent line, a surface by its tangent plane. This approach simply will not work on a highly irregular curve or surface.

Nevertheless, we can imagine highly irregular curves. Originally these were seen as 'pathological' objects whose purpose was to exhibit the limitations of analysis. They were counter-examples, serving to remind us that the capacity of mathematics for nastiness is unbounded. The pure mathematician's motto is Murphy's Law: 'Anything that can go wrong, will go wrong.' And the wise mathematician or scientist always wants to know what can go wrong. Often this is a starting-point for finding new ways for things to go right.

For example, during the eighteenth and nineteenth centuries it was widely assumed that any continuous curve must have a well-defined tangent (that is, any continuously varying quantity must have a well-defined instantaneous rate of change) at 'almost' any point. The only exceptions were the corners, where the curve makes an abrupt change of direction. However, in a lecture to the Berlin Academy in 1872, Karl Weierstrass showed that this is not true. It is, in fact, about as false as it is possible to get. He described a class of curves that are continuous, but have no points where the tangent is well defined. The basic idea is to add together infinitely many increasingly tiny 'wiggles'. The resulting curve is continuous – it has no gaps – but it wiggles so rapidly that there is no sensible way to construct a tangent. Anywhere.

Again, in 1890 Giuseppe Peano constructed a curve that passes through every point of the interior of a unit square. This curve demonstrated the complete inadequacy of the common idea of 'dimension' as the number of (continuously varying) parameters needed to specify a point. Peano's curve takes a square, with its two dimensions and standard parametrization by two coordinates (north–south and east–west), and reparametrizes it by a single variable: how far you have to go along Peano's curve in order to hit a given point.

In 1906 Helge von Koch gave an example of a curve of infinite length that bounds a finite area: the snowflake. (Fig. 1.1) It is constructed by starting with an equilateral triangle, and erecting on each side a smaller triangle, one-third the size. This construction is repeated to infinity. Like Weierstrass's curve, the snowflake is continuous but has no tangent. A similar repetitive process occurs in the construction of one of the simplest and most fundamental pathological sets of all: the Cantor Set, named for Georg Cantor who used it in 1883 (although it was known to Henry Smith in 1875). It is constructed by repeated deletions of the middle third of an interval. (Fig. 1.2)

The mathematical community – even leading figures – found it hard to come to terms with these unsettling discoveries. Henri Poincaré dismissed

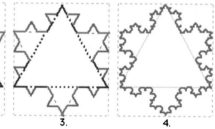

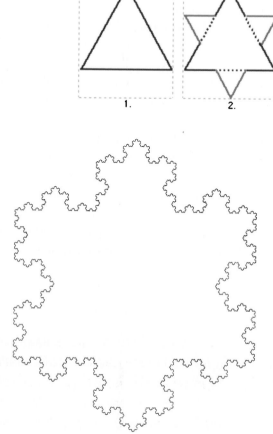

Fig. 1.1 To a casual observer this is a snowflake, but mathematically it is a classic fractal shape, constructed out of one equilateral triangle, with the middle third of each side removed and new equilateral triangles drawn out to the edge, their middle third removed, smaller triangles drawn out in turn, and so on.

Fig. 1.2 The Cantor Set: first developed in 1883, it is constructed by repeated deletions of the middle third of an interval.

them as 'a gallery of monsters', and Charles Hermite deplored what he called a 'lamentable plague of functions with no derivatives'. More recently Jean Dieudonné wrote: 'Some mathematical objects, like the Peano curve, are totally non-intuitive ... extravagant.' But Dieudonné was not suggesting they lacked interest, just that they were difficult to wrap your head round.

It is only fair to add that the undue proliferation of such sets, without any clear purpose in mind, can easily become an exercise in futility. So Poincaré and Hermite did have some basis for their complaints. But as time passed, most mathematicians came to accept that these sets play a legitimate, indeed crucial, role in mathematics: they demonstrate that there are limits to the applicability of classical analysis. In fact, this realization stimulated the development of new kinds of non-classical analysis, which turned out to be important in their own right. Indeed by 1900 the great German mathematician David Hilbert could refer to the whole area as a 'paradise' without causing ructions. Nonetheless, many mathematicians were perfectly prepared to operate within the classical limits. They saw the 'pathologies' as 'artificial' objects, unlikely to be of any importance in the study of Nature.

Nature, however, had other ideas.

How long is the coast of Britain?

The fractal geometry of coastlines

One of the formative examples of fractals is the geometry of coastlines. In particular: how long is a coastline? Coastlines are notoriously irregular, and the answer to the question depends on how the measurement is made. The simplest method is to take a fixed finite length x and move along the coastline in steps of length x. Adding these steps together gives a total length $L(x)$. (Fig. 1.3)

Fig. 1.3 Mapping a coastline: the actual length depends on how many steps of length x one takes. If $x = 1$ km the length will be considerably less than if the length were 1 m; and this will be far less than steps of 1 cm; and so to infinity.

If the coastline is smooth, in the rigorous mathematical sense, then when x is small enough, the coastline is very close to a straight line. For a straight line, the value of $L(x)$ tends to a definite limit L as x tends to zero, and that limit is the length of the straight line in the usual sense. It follows that if the coastline is a smooth curve, $L(x)$ also tends to a definite limit L as x tends to zero, and that limit is the length of the curve in the usual sense. In other words, if x is small enough, $L(x)$ is an approximation to the total length that is close enough on the scale of the model chosen.

What actually happens, with real coastlines, is quite different. Small bays of diameter smaller than x are missed by the stepping procedure. Although reducing the value of x must in some sense improve

One of the formative examples fractals is the geometry of coastlin In particular: how long is a coastlir Coastlines are notoriously irregu and the answer to the question deper on how the measurement is ma

the approximation, by 'noticing' ever smaller bays, there will still remain irregularities on some scale smaller than x, at least until we get down to molecular proportions where the whole exercise becomes meaningless. Because coastlines are fractal, the value of L(x) grows without limit, and the length is infinite.

In the absence of a finite limiting value, it is often useful to study how a quantity tends to infinity. Is the growth rate fast and explosive, or slow and steady? In other words, what is the 'asymptotic' (when a curve tends towards but never reaches a straight line) behaviour? Lewis Fry Richardson once made an empirical study of the asymptotic problem, for real coastlines, and found an excellent empirical law: $L(x) \sim kx^{1-D}$ for certain constants k and D. The value of D is much the same for most coastlines on planet Earth, presumably for geological reasons, and in particular $D \sim 1.25$ for the coast of Britain.

To gain an intuitive feeling for what this result means, compare Britain to a snowflake curve. The construction of the snowflake is too regular to correspond to a real coastline, but as far as the main feature – structure on all scales – goes, it's not bad.

For simplicity, measure its length using values

$x = 1, \frac{1}{3}, \frac{1}{9}, \frac{1}{27}$, and so on.

Then $L(1) = 1$, $L(\frac{1}{3}) = \frac{4}{3}$, $L(\frac{1}{9}) = (\frac{4}{3})^2$, $L(\frac{1}{27}) = (\frac{4}{3})^3$, and so on.

In general $L((\frac{1}{3})^n) = (\frac{4}{3})^n$. Let $x = (\frac{1}{3})^n$, and note that $\frac{4}{3} = (\frac{1}{3})^{1-D}$ where $D = \frac{\log 4}{\log 3}$.

Then $L(x) = x^{1-D}$ and $D = 1.2618$.

This is very close to the empirical estimate $D = 1.25$ for the coastline of Britain.

I am not claiming that this implies that Britain is a snowflake. The snowflake curve's geometry is much too regular. Nevertheless, we may interpret the above calculation in the following terms. Suppose a real coastline has the same statistical distribution of bays and promontories, sub-bays and subpromontories, as does the snowflake curve. Then the value of L(x)

should follow the same asymptotic law as for the snowflake, and thus lead to the same D. If the statistical distribution is similar to that of the snowflake, but not quite the same, then the constant D should change slightly. So we conclude that the coastline of Britain has pretty much the same 'roughness' as the snowflake – but is maybe just a tad smoother.

The combinatorial regularity of the snowflake is essentially a scaling law. If a small section of the curve is suitably magnified, then it looks exactly like some larger section of the original. The constant D describes, in a quantitative manner, the precise scaling required. Here, if four copies of a segment of the curve are suitably assembled, the result has exactly the same shape as the segment, but is three times as large. The value $\frac{\log 4}{\log 3}$ of D is built from those two numbers. This property is called self-similarity. The same idea holds for coastlines, but now the scaling affects the statistics, not the curve as such. Instead of asking that a magnified version of a section of coastline should be exactly the same as the original, we ask that it should be a plausible picture of a coastline on the same scale as the original. Or, to put it another way: if you are presented with a map of a coastline, without any other markings and with no indication of the scale, then there will be no way to determine the scale just by studying the map.

Innumerable other natural phenomena exhibit structure on a wide range of scales, connected by suitable scaling laws. For instance, the bark of a tree, the ripples on the ocean, vortices in a turbulent fluid, landscapes, the inner surface of the lung, the holes in a sponge, the surface of a soap flake. Therefore we expect there to be some regime of mathematical modelling in which the 'pathological' curves and surfaces that were so despised by the classical mathematicians find natural application to the real world. Since scaling laws appear to be fundamental to the whole enterprise, the initial emphasis should be on understanding what they have to tell us. And the first thing they tell us extends the usual notion of 'dimension' in a radical way.

Fractal dimension

It turns out that the number D introduced above may be interpreted as a dimension. This may seem a rather curious idea, since the usual notion of dimension is always a whole number, but there are plenty of precedents in mathematics. The concept 'number', for example, originated in counting – one sheep, two sheep, three sheep. In this context, half a sheep makes no sense. But in the butcher's shop – or, less grimly, at the moneylender's, where a person might own a half share in a sheep – the extension of the number system to fractions is natural. Again, we are used to the idea that the nth power of a number is obtained by multiplying n copies of that number – so that the fifth power of 3, for example, is $3^5 = 3 \times 3 \times 3 \times 3 \times 3 = 243$. What, then, is the halfth power? What you get by multiplying half a copy of a number by itself? That makes little sense, but the halfth power makes excellent sense: it is the square root. Multiply the halfth power by itself, and you get back the first power – the original number. Twice a half is one – easy.

In fact, the generalization of dimension that occurs in fractal geometry is reasonable from several points of view. To see why, we begin by reviewing the usual concept of dimension. (Fig. 1.4)

(a) A line segment has dimension 1, by which we mean that any point in the segment can be specified using just one coordinate, one number. The point x lies x units to the right of the left-hand end of the segment.

(b) A square has dimension 2, by which we mean that any point in the square can be specified using just two coordinates (x, y). Here x is the distance from the left-hand edge and y is the distance from the bottom edge.

Fig. 1.4 The concept of dimension in geometry: (a) a line has 1 dimension and 1 coordinate; (b) a square has 2 dimensions and 2 coordinates; (c) a cube has 3 dimensions and 3 coordinates.

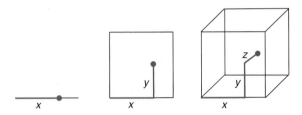

(c) A cube has dimension 3, by which we mean that any point in the cube can be specified using just three coordinates (x, y, z). Here x is the distance from the left-hand face, y is the distance from the bottom face, and z is the distance from the back face.

In these examples, the dimension of the object is the number of independent directions in space that it occupies. No directions are needed for a point, so it has dimension 0. A line lies along one direction, a square lies in two (a plane), whereas a cube requires three. Similar ideas apply to curved lines and surfaces. A curve has dimension 1. The surface of a smooth object, such as a sphere or torus, has dimension 2. A solid object, such as a solid sphere or a solid torus, has dimension 3. This concept of dimension is always a whole number. A point has dimension 0, a curve has dimension 1, a surface has dimension 2, a solid has dimension 3. With a suitable act of imagination, we can go into spaces of dimension 4, 5, 6, and so on – see Abbott (1884) and its modern sequel Stewart (2001). Engineers will recognize this concept as the number of 'degrees of freedom' of a system – the number of coordinates needed to determine its state – so that space-time, with 3 space coordinates and one time coordinate, is 4-dimensional.

The dimension of even a simple system can be surprisingly large. For example, describing the position and velocity of the Moon in space requires six numbers: three position coordinates, and three components of velocity relative to those coordinates. So the 3-body system composed of the Earth, Moon, and Sun, which is basic to astronomy, is an 18-dimensional system. Each body requires 3 coordinates of position in space and a further 3 of velocity.

A more extreme case is something we all carry around with us: the human body, with its innumerable flexible joints. Look at your hand. Each finger can be bent

Above: A solar eclipse

At a conservative estimate, the 'configuration space' for the human body – the totality of possible shapes into which it can be bent – is at least 101-dimensional. Yes, we live in space of 3 dimensions, and a space-time of 4, but the complete range of possible configurations of the human body forms

through some angle, and those angles are pretty much independent of each other. So just to describe the state of your hand, you need a 5-dimensional space of possible configurations. In fact, fingers can bend sideways (a bit) too, so 10 dimensions is a more realistic number. Your two hands and two feet now require at least 40 dimensions to capture all possible combinations of positions, and then there are your wrists, elbows, shoulders, ankles, knees, thighs ... and your head, eyelids, and waist.

At a conservative estimate, the 'configuration space' for the human body – the totality of possible shapes into which it can be bent – is at least 101-dimensional. Yes, we live in space of 3 dimensions, and a space-time of 4, but the complete range of possible configurations of the human body forms a conceptual 'space' with 101 dimensions.

This notion is called topological dimension because shapes that can be continuously deformed into each other have the same dimension. Thus a wiggly curve has the same dimension, 1, as a straight line; a wobbly surface has the same dimension, 2, as a plane. And if a shape is magnified by some scale factor – say tripled in size – then its dimension remains unchanged.

Scaling laws are more sensitive: they involve not just shape, but size. Distances are important, scale matters. What count are not topological properties, but metric ones. This extra ingredient opens up the possibility of finding an extended notion of dimension which

(a) agrees with the usual definition for smooth curves and surfaces;

(b) applies to more general spaces, such as the snowflake or the Cantor Set; and

(c) reflects metric, not topological, properties, especially behaviour under scaling.

The price we pay for such an extension, however, is that the resulting concept of dimension is forced to take non-integer values. It turns out to be a price well worth paying – imaginative ideas that take us out of our comfortable world usually are.

The simplest such generalization (there are many) is the similarity dimension. This concept is based on scaling properties; it is a little too special to be entirely satisfactory, but when it does work it is very easy to understand.

Consider a unit square. If its sides are divided into n equal parts, then it can be cut into $N = n^2$ subsquares, each similar to the original. With a similar dissection of a cube, we find that $N = n^3$; with a 4-dimensional hypercube we get $N = n^4$. And with a homely line segment, $N = n^1$. (Fig. 1.5)

The pattern is obvious: if the dimension is d, then $N = n^d$. Taking logarithms and solving, we get $d = \frac{\log N}{\log n}$. All perfectly reasonable, and equivalent to standard geometrical properties of these simple shapes.

So let's try a shape that is not quite so simple: the archetypal 'pathological set', the Cantor Set. Remember: to form a Cantor Set, start with a line segment, remove its middle third to get two segments each one-third the size; then repeat indefinitely. What's left is the Cantor Set. It is clear that after the initial step, we construct two separate Cantor Sets, each one third the size of the whole; the Cantor Set itself is obtained by uniting these two subsets. In other words, the Cantor Set can be broken into two pieces (N = 2) each one third as big (n = 3). By formal analogy, the dimension of the Cantor Set 'ought' to be $d = \frac{\log 2}{\log 3} = 0.6309$, which is not a whole number. This may seem curious, but it makes a lot of sense because:

(a) it accurately reflects the scaling properties of the Cantor Set: two copies make a set just the same shape but three times as big; and

(b) the dimension is intermediate between 0, the dimension of a finite set of points, and 1, the dimension of a curve. This agrees with the intuitive idea that the Cantor Set is rather less than a curve, since it has gaps, but is more closely clustered than a finite set of points.

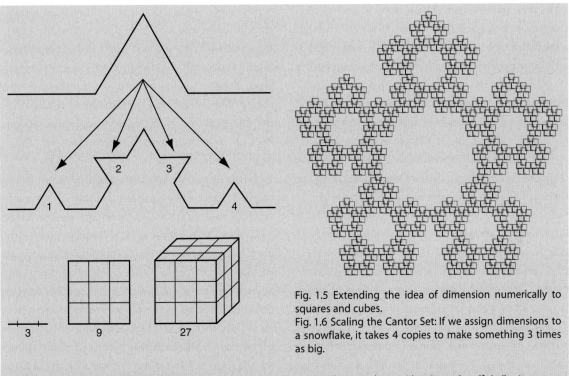

Fig. 1.5 Extending the idea of dimension numerically to squares and cubes.
Fig. 1.6 Scaling the Cantor Set: If we assign dimensions to a snowflake, it takes 4 copies to make something 3 times as big.

The main restriction in the definition of the scaling dimension is that it requires the set under consideration to the self-similar. It must coincide with small pieces of itself when suitably scaled.

Similar reasoning lets us assign a dimension to the snowflake. If we work (for convenience) with one segment of the snowflake, we see that it takes 4 copies to make something 3 times as big. (Fig. 1.6) Thus N = 4, n = 3, and $d = {\log 4}/{\log 3} = 1.2618$. Recognize this number? It is the constant D that appeared in the coastline calculation for the snowflake. That is, we have interpreted D as a scaling dimension. The same kind of game can be played with all scaling laws. (This, by the way, is why we used 1−D in the formula, instead of just D.)

For the snowflake, d lies between 1 and 2, and this again agrees with visual intuition. The snowflake is 'more than' a smooth curve, but hardly constitutes a surface.

The main restriction in the definition of the scaling dimension is that it requires the set under consideration to be self-similar. It must coincide with small pieces of itself when suitably scaled. More general notions of dimension also exist. For theoretical work, the best is the Hausdorff-Besicovitch dimension, introduced in 1919 and generalized in 1929. Related but different concepts of dimension are often used in experimental work because they are easier to measure; see Falconer (1990).

Brownian motion

Particle paths

One potential application for a theory of 'pathological' curves has been around since 1828. It demonstrates that the common assumption that the only curves needed to do physics are smooth ones is completely wrong.

In that year, Robert Brown drew attention to a curious phenomenon that he had observed through a microscope. If very small particles are suspended in a fluid, apparently at rest, then on close inspection it will be seen that they undergo frequent irregular motions. Brown suggested that this movement might be caused by the motion of molecules in the fluid, brought about by heat. In 1926 Jean Perrin wrote:

The direction of the straight line joining the positions occupied at two instants very close in time is found to vary absolutely irregularly as the time between the two instants is decreased. An unprejudiced observer would therefore conclude that he is dealing with a function without derivative, instead of a curve to which a tangent could be drawn ... At certain scales and for certain methods of investigation, many phenomena may be represented by regular continuous functions. If, to go further, we attribute to matter the infinitely granular structure that is in the spirit of atomic theory, our power to apply to reality the rigorous mathematical concept of continuity will greatly decrease.

Perrin had discovered that the 'pathological' curves that both intrigued and repelled the mathematicians were entirely natural, in the sense that they had sensible counterparts in nature; they arose in areas as fundamental as molecular motion. He carried out experiments on these ideas, and was awarded the Nobel prize in 1926.

In the 1930s Norbert Weiner formulated a mathematical model of Brownian motion, showing that it possesses exactly this feature of non-differentiable particle trajectories. Weiner's work, roughly speaking, is carried out in the context of a random function of time, such that the probability of moving any given distance at any instant is determined by the classic 'bell curve' or Gaussian distribution, while the direction is completely and uniformly random. In such a setting, particle trajectories are almost nowhere differentiable.

Within the past few years, methods from fractal geometry have led to major advances in our mathematical understanding of Brownian motion. In his 1982 book, Mandelbrot made a conjecture about the fractal dimension of a typical particle trajectory. At the time, his evidence was derived mainly from computer simulations. Imagine a single particle obeying Brownian motion in the plane. Follow its trajectory for a fixed period of time, obtaining an extremely wiggly curve. Mandelbrot conjectured that the fractal dimension of this curve – more exactly, of those parts of the curve that lie on the outer edge of the shape that it forms – is exactly $4/3$.

In 2000, Gregory Lawler, Oded Schramm, and Wendelin Werner announced a rigorous proof of this conjecture, together with several other fractal properties of Brownian motion. Their proof involves mathematical analogies with other fractal processes. Other mathematical physicists have suggested different proofs based on links with the theory of quantum gravity, which had not previously been related to Brownian motion. (The relation is mathematical:

In that year [1828], Robert Brown drew attention to a curious phenomenon that he had observed through a microscope. If very small particles are suspended in a fluid, apparently at rest, then on close inspection it will be seen that they undergo frequent irregular motions. Brown suggested that this movement might be caused by the motion of molecules in the fluid, brought about by heat.

Innumerable other natural phenomena exhibit structure on a wide range of scales, connected by suitable scaling laws. For instance, the bark of a tree, the ripples on the ocean, vortices in a turbulent fluid, landscapes, the inner surface of the lung, the holes in a sponge, the surface of a soap flake.

the physical interpretations of the two areas are quite different.) Lawler and co-workers also proved that the fractal dimension of the set of 'cut points' of the curve is $^3/_4$ and that of the set of 'pioneer points' is $^7/_4$. A cut point is one whose removal causes the curve to fall apart into two disconnected pieces. A pioneer point is one that finds itself on the outer boundary at the instant the curve first reaches that point. These results illustrate the high level of mathematical detail in our new understanding of particle paths in Brownian motion, obtained through fractal geometry.

Turbulence

One of the most important and baffling phenomena in fluid dynamics is turbulence: irregular, twisting flow-patterns far removed from the smooth 'laminar' flows beloved of the classical analysts. Until recently, turbulence has been studied by a variety of ad hoc analytical methods and probabilistic models, but relatively little attention has been paid to the geometry of turbulence. Yet the geometry contains hints of a deeper structure that the analytic approach misses. Turbulence involves motion on a wide range of scales, large and small. As Lewis Fry Richardson put it in 1922:

> Big whorls have little whorls,
> Which feed on their velocity;
> And little whorls have lesser whorls,
> And so on to viscosity.

Could fractals be involved in the geometry of turbulence?

This suggestion was made by Mandelbrot in about 1960. It re-emerged in a very different guise from the topological dynamics of the mid-1970s, and it now appears to be firmly established by careful experiments using a variety of 'small-scale' laboratory systems. Of course the theory met with fierce resistance along the way, as occurs with anything genuinely new in science, especially when it is advocated by interlopers from another field. To be fair, these experiments establish the occurrence of fractal geometry in weak turbulence; fully developed turbulence is quite another matter – but if anything, that looks even more fractal.

Turbulence may be confined to certain regions of an otherwise smooth flow, or it may appear suddenly everywhere. It can appear and disappear intermittently. In the Taylor vortex experiment, where fluid is observed in the region between two concentric rotating cylinders, spiral turbulence occurs in patches on a predominantly helical flow like a spinning barber's pole. The boundaries of turbulent regions typically have a complex local structure: billows upon billows, whorls upon whorls. The region around Jupiter's Great Red Spot is typical of such behaviour.

The topological approach to turbulence was initiated by Ruelle and Takens (1971), who suggested a scenario for the transition to turbulence in terms of the creation of a so-called 'fractal attractor' in the dynamics. Harry Swinney, Jerry Gollub, and others carried out experiments using lasers to measure the speed of the fluid, and confirmed the general conceptual framework, though not the precise scenario originally proposed.

In larger systems, the transition to turbulence is a much more complex affair. So we still have much to learn about turbulence. Fractal geometry can help us make advances, but it cannot answer everything. What can?

Fractal drums

Fractals as vibrational modes

In 1996 Michael Lapidus and colleagues studied the vibrational modes of a drum shaped like the snowflake curve. The practical spin-off from such research includes a better understanding of why a rocky coastline dissipates rough seas better than a smooth one – important for coastal defences, and one reason why all the old straight-sided promenades are being replaced by irregular heaps of spiky concrete. It also helps to explain why the foliage of trees proves so resistant to the wind, and how our system of elastic-walled veins, arteries, and capillaries absorbs the thud of a beating heart with surprisingly little damage. It offers new insights into how radar waves bounce around in mountainous terrain, and how laser beams might reflect from the cratered landscape of the Moon.

To understand what a vibrational mode is, think about a guitar. Pluck an open string: it produces a single note, the fundamental. Now rest your finger gently against the exact middle of the string, pluck again, and quickly lift your finger off. You hear a high-pitched ping!, exactly one octave higher up the musical scale than the fundamental. If you place your finger one third of the way along the string, you can create an even higher note, and so on. Your finger is selecting various vibrational modes of the string. When the string sounds its fundamental, it forms a single standing wave. The ends are fixed, but the rest of the string moves up and down in a regular, repetitive fashion. For the octave, two such waves fit into the length of the string, and when one goes up, the other goes down. In between is a fixed node, the place where you put your finger. A guitar string can – in theory – vibrate with any whole number of waves. So as well as the fundamental frequency, it has vibrational frequencies that are twice the fundamental, three times, four times, and so on. Its spectrum, its list of possible frequencies, consists of all whole number multiples of the fundamental.

Every shape has an acoustic spectrum. Physically, you can observe the spectrum by making the shape from metal and hitting it – or anything equivalent, such as making the shape from a soap film and watching it wobble, or carving the shape as a cavity in a lump of metal and filling it with microwaves. Mathematically, the most significant aspect of the vibration is the list of frequencies of the natural vibrational modes, called the 'spectrum' of the shape.

In 1910 the physicist Hendrik Lorentz lectured on the spectrum of electromagnetic waves in an enclosed cavity, which is the same mathematical problem in a different physical realization, and made a bold conjecture. Suppose you choose a shape, and arrange all the frequencies of its spectrum in ascending numerical order. Now ask how fast those numbers grow as you run along the list. For simplicity, consider a two-dimensional cavity – an area in the plane. Then, said Lorentz, no matter what shape the cavity may be, the frequencies are approximately $\frac{2\pi}{A}$, $\frac{4\pi}{A}$, $\frac{6\pi}{A}$, $\frac{8\pi}{A}$, $\frac{10\pi}{A}$, and so on, where A is the area of the cavity and π is the usual 'pi' that we all know and love. Moreover, the approximation gets better and better the higher the frequencies become.

David Hilbert, the world leader in mathematics at the time, attended the lecture and was really impressed by Lorentz's conjecture, but he allegedly said that he didn't expect to see a proof in his lifetime. If so, he was unduly pessimistic: less than two years later his former student Hermann Weyl proved something far more general, using a technique – integral equations – that he had learned from his master, Hilbert. Weyl's ingenious argument showed that Lorentz's conjecture is valid not just in the plane, but in space of n dimensions, for any n. But now, instead of the area, you must use the multidimensional equivalent of volume; the constant π must be replaced by a more complicated expression related to the unit n-dimensional sphere; and the frequency must be replaced by its $\frac{n}{2}$th power. Nonetheless, the main point remains: the frequencies in the spectrum are related to the volume of the object, so in particular if the object gets smaller, the frequencies increase.

It is here that fractal drums get in on the act. Weyl originally proved his result for objects with a fairly smooth boundary, but over the years his formula was extended to objects with fractal boundaries – such as the snowflake curve. In 1979 the physicist Michael Berry was thinking about light scattering from

... there is a curious consequence: you can 'hear' the fractal dimension of the boundary of a drum – that is, you can determine it from the drum's spectrum. Ideas of this kind go back to a famous article written by Mark Kac in 1966, called 'Can one hear the shape of a drum?'

irregular surfaces, and he came up with a conjectured improvement to Weyl's formula. This improvement involved an extra term, proportional to the fractal dimension of the object's boundary. According to Berry, the extra term in Weyl's formula ought to be proportional to the frequency raised to the power of half the fractal dimension of the boundary.

If so, there is a curious consequence: you can 'hear' the fractal dimension of the boundary of a drum – that is, you can determine it from the drum's spectrum. Ideas of this kind go back to a famous article written by Mark Kac in 1966, called 'Can you hear the shape of a drum?' Kac pointed out that Weyl's formula shows that you can hear the area of a drum, no matter what the shape of the drum's rim might be, and asked what else can you hear? He set the ball rolling by proving that the spectrum also determines the drum's perimeter, and even its connectivity – how many holes it has. It turned out that there are some features that you can't always hear. Various mathematicians found examples where two different high-dimensional shapes had the same spectrum, meaning that their shape cannot be 'heard' in complete detail. The first example was in 16 dimensions. By 1982 the dimension was down to 4, and in 1992 Carolyn Gordon, David Webb, and Scott Wolpert knocked the problem on the head

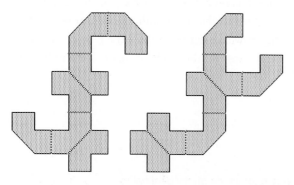

Fig. 1.7 Two 2-dimensional drums: how to share the same spectrum but with different shapes.

by finding two 2-dimensional drums with the same spectrum, but different shapes. (Fig. 1.7)

Even though you can't hear everything you'd like to, the question 'what can you hear from an object's spectrum?' is important. There are many cases where it is impossible to observe an object directly, but far easier to observe its vibrations. A good example is seismology, which infers the inner structure of the Earth from vibrations generated by earthquakes. Helioseismology does the same for the Sun. Oil companies use sound waves created by surface explosions to look for oil deposits deep underground. Children rattle wrapped presents to try to work out what's inside. Berry's proposal, if correct, would have added fractal dimension – roughness – to the list of hearable quantities. But it was not to be: in 1986 J. Brossard and R. Carmona found a shape for which the correction term was not related to fractal dimension in the anticipated manner.

All was not lost, however: they suggested that a different measure of roughness, the less familiar 'Minkowski dimension', might work instead. A version of this revised conjecture was proved by Michel Lapidus and Jacqueline Fleckinger-Pellé in 1988. The most recent work on the spectra of fractals has gone beyond general results on the distribution of frequencies, to look in detail at the actual vibrational patterns. Since 1989 the experimentalists G. Sapoval, Th. Gobron and A. Margolina have been studying mechanical vibrations of fractal objects – such as a steel plate etched with a laser to produce a 'squareflake'-shaped groove – a variant of the snowflake curve made from squares rather than triangles. Sapoval's team discovered new effects in vibrating fractals, effects that don't occur for more traditional shapes. For example, it looks as though wave motion in regions with fractal boundaries can be localized – small regions of the object vibrate noticeably, but the rest hardly moves.

In 1995 Lapidus and M. Pang performed a rigorous mathematical analysis of the fundamental mode, the lowest frequency vibration, of the snowflake curve. They discovered that for this mode, the vibrating 'drumskin' can become infinitely steep near certain points of the boundary – mainly those where obtuse angles (greater than a right angle) occur. In a physical analogy, the vibrating fractal drum experiences infinite stress at such points. The rigorous mathematics confirms an effect observed in earlier experiments. What does a vibrating fractal really look like? Using an Onyx computer from Silicon Graphics, and some cunning numerical methods, Lapidus's team has drawn the first fifty snowflake harmonics – the vibrational modes of a drumskin whose boundary is the snowflake curve. Their findings confirm previous discoveries, and add some new puzzles.

In the fundamental mode, as predicted by general theory, the whole drumskin moves upwards or downwards at the same time – there are no nodal curves where the drumskin is stationary, except at the boundary. The extremely steep gradient near obtuse-angle boundary points is clearly visible. Conversely, it turns out that the gradient near acute angles is zero – the membrane is flat, the stress is zero.

The spectra of fractals comprise a vibrant new area of science – pun intended – with the potential to solve a lot of puzzles, including many of the questions raised earlier. For example, think about waves hitting a rocky coast. A coastline is much better

leaves and branches of trees
fractal, and this may be why
make better barriers to the
than a simple flat fence –
ct that modern agriculture
arting to rediscover after
ades of rooting out
ges to produce vast,
windswept, fields.

approximated by a fractal than by a smooth curve, so the action of ocean waves on a rocky coast should be much closer to the snowflake model than to classical analyses of waves hitting a flat boundary. A flat boundary reflects most of the waves' energy back out to sea, but a fractal boundary seems able to absorb wave energy, or isolate it into small patches.

The leaves and branches of trees are fractal, and this may be why they make better barriers to the wind than a simple flat fence – a fact that modern agriculture is starting to rediscover after decades of rooting out hedges to produce vast, but windswept, fields.

Similarly the human circulatory system, with its repeated branching pattern of ever-smaller veins and arteries, is fractal. As well as conducting blood effectively to all parts of our bodies, this fractal structure may also help the blood vessels to absorb and dissipate the stress caused by the heavy thumps of a beating heart.

Fractals in technology

Practical Applications

Although the main importance of fractal geometry is as a scientific tool, most science eventually acquires practical applications, and fractal geometry is no exception. In fact, the next few decades will probably see an explosion of fractal-based technology. Already the applications are quite diverse. Here are just a few examples.

Nearly all machines include springs. A video recorder, for example, contains several hundred of them. Springs are made by coiling wire on special machines. Until recently, a big problem was to test in advance whether wire would make good coils or poor ones. The only method was to try them and see, which could take a day or more. Now a test, based on fractal geometry, takes only two minutes. It is embodied in the FRACMAT machine (short

for 'fractal materials') and was invented by the Institute of Spring Technology in Sheffield, together with a team from the University of Warwick. The idea is to make a long test coil on a metal rod, and then analyse the fractal patterns in the spacings of the coils. The type of fractal structure present correlates closely with the desirable quality of 'coilability'.

Mines are dangerous places to be. Earthquakes or other stresses building up in the walls of a mine can sometimes lead to devastating 'rockbursts', often with fatal consequences for miners nearby. In 2000

…the combined sounds of thousands of fractures is characteristic of the rock's behaviour.

engineers at the Southwest Research Institute, San Antonio, discovered that rockbursts can be detected as they build up by listening to the sounds made by the rocks. Every tiny fracture makes its own popping sound, and the combined sounds of thousands of fractures is characteristic of the rock's behaviour. Simon Hsiung and colleagues at SWRI realized that the pattern of cracks is fractal, and therefore the associated sounds should also be fractal. The fractal dimension of the cracks (and the associated sounds) first grows steadily; then it suddenly begins to drop. Soon after that point is reached, the rocks blow apart. In small-scale experiments the warning time is a few minutes, but in the large-scale cracking that occurs in a real mine it may be days. If so, miners can either evacuate an area where a rockburst is due to occur, or take steps to relieve the stress so that it never happens.

One of the main limitations on the effectiveness of mobile phones has been the antenna that receives and transmits radio signals. The first radio receivers used little more than a bare wire as an antenna, and until recently most mobile phones were little more sophisticated. Then composite antennas, made of thousands of smaller ones, became available. They are usually arranged either regularly in a rectangular grid, or at random. In 1999 Dwight Jaggard and Douglas Werner discovered that a fractal arrangement of these micro-antennas combined the robustness of a random array with the efficiency of a grid. Already there is some theoretical understanding of why fractal shapes work so well. Nathan Cohen and Robert Hohfeld proved that if an antenna is to work equally well at any frequency, then it must have two features: symmetry, and self-similarity. Many fractals, for example the Sierpinski gasket (Fig. 1.8), have both.

Digital communications, be they television or computers, transmit visual images as a sequence of binary digits 0 and 1. The easy way to turn an image into such a sequence is to read off the black

Fig. 1.8 The Sierpinski gasket: this nest of triangles displays two of the common characteristics of fractals – symmetry and self-similarity.

and white 'pixels' – tiny picture elements – from a regular grid, with 0 representing white and 1 black. More complex codes can represent shades of grey or colours. The resulting list of digits is huge, as anyone who uses a scanner knows. Engineers are always looking for ways to encode the sequences so that the same image can be represented by fewer digits. For example, in photographs there is often a lot of blue sky, so using a short code for 'blue' makes more sense than a longer code. In video, the most important feature is which pixels change from one frame to the next; the rest can be left the same as they were. And so on.

In 1996 Iterated Systems of Atlanta, a company founded by mathematician Michael Barnsley (see the chapter in this book written by Michael with his wife Louise), developed a data-compression system for video images based on fractal self-similarity. Roughly speaking, a computer compares small regions of the image with larger ones, and lists cases where the two have much the same form. From this list, which is typically much shorter than a grid of pixel codes, it is possible to reconstruct the image with almost perfect accuracy.

Above: The nebula

Through fractal eyes

A Voyage of discovery

What, then, have fractals taught us? Until recently it was fair to say, as I did in From Here to Infinity in 1996, that: 'The contribution of fractals to our understanding of the natural world is not so much one of technology as of what used to be called natural philosophy.' Fractals provide a unified point of view on certain kinds of complexity and irregularity in the natural world, and open a path for a mathematical attack. They act as an organizing principle, not as a computational tool like calculus or linear algebra or numerical methods.

Nowadays it must hastily be added that the computational and technological aspects of the subject are advancing rapidly, as more and more scientists wake up to the new methods, stop trying to dismiss them, and start trying to use them. The recognition of fractals as basic geometric forms, amenable to analysis but having quite different characteristics from the familiar smooth forms such as spheres and cylinders, opens our eyes to a new range of phenomena and sensitizes us to new points of view. Instead of being seen as 'erroneous' or 'uninteresting', and hence avoided, these phenomena are seen as something to be deliberately sought out and understood.

One measure of the maturity of a mathematical theory is the extent to which it studies things that a previous generation dismissed as being 'special', 'pathological', 'non-generic', 'accidental', 'abnormal', 'coincidental'... There are many words in the language for 'we don't understand this', and science uses them all. Major new theories often arise when someone takes such unorthodoxies seriously, and investigates them in their own right with an open mind.

Examples are ready to hand. When Edward Lorenz (1963) first observed irregular solutions to a model of the weather, hardly anybody took any notice, yet today there is scarcely any branch of science that does not make contact with Chaos Theory, which grew from that discovery (and several others). Again, the study of chemical oscillators, now highly fashionable, was for many years thought to be akin to the search for a perpetual motion machine. The list is virtually endless: the only counterbalance to this memorial to human folly, narrow-mindedness, and prejudice is the even more extensive list of comparably unorthodox ideas that have proved utterly worthless. Not all novelty is valuable, and being enthusiastic does not guarantee being right.

As an identifiable area of mathematics with its own characteristic point of view and body of techniques, fractal geometry has now 'arrived'. Its viewpoint is recognizable and recognized. But for all the beauty of its pictures and the breadth of its vision of the natural world, more work needs to be done before the theory becomes fully established. Many of its models are descriptive rather than explanatory. Fractal 'fake' mountains look like real ones, but we have as yet little understanding of how erosion processes produce the fractal structure. We often cannot compute the fractal dimension, let alone anything more sophisticated, from basic physical principles.

As I said, all this is starting to change, as the theory moves into its next phase. But even a purely descriptive theory adds scientific value. If cellular tissue is best modelled by a fractal, then there is no point in treating it as a rectangular slab. Today, the role of fractal geometry is not primarily to add new weapons to the pure mathematician's armoury

(though it sometimes does), or to help us make a better mobile phone (though it sometimes does), or to provide insight into the structure of the universe (though it sometimes does). It is to open our eyes to an entire realm of mathematics. In 1996 I put it this way: '[Fractals] are important because they suggest that, out there in the jungle of the unknown, is a whole new area of mathematics, directly relevant to the study of nature.' We have now explored enough of the jungle to realize that what remains to be explored is even bigger, and more exciting, than we ever imagined.

References

E.A. Abbott (1884). Flatland: a Romance of Many Dimensions. Seeley, London.

K. Falconer (1990). Fractal Geometry. Wiley, Chichester.

M. Kac (1966). Can one hear the shape of a drum? American Mathematical Monthly 73 1–23.

M.L. Lapidus, J.W. Neuberger, R.J. Renka, and C.A. Griffith (1996). Snowflake harmonics and computer graphics: numerical computation of spectra on fractal drums. International Journal of Bifurcation and Chaos 6 1185–1210.

G. Lawler, O. Schramm, and W. Werner (2000). The dimension of the planar Brownian frontier is $4/3$. Preprint, Duke University.

E. Lorenz (1963). Deterministic nonperiodic flow. Journal of the Atmospheric Sciences 20 130–141.

B. Mandelbrot (1977). Fractals: Form, Chance, and Dimension. Freeman, San Francisco.

B. Mandelbrot (1982). The Fractal Geometry of Nature. Freeman, San Francisco.

D. Ruelle and F. Takens (1971). On the nature of turbulence, Communications in Mathematical Physics 20 167–192.

I. Stewart (1996). From Here to Infinity. OUP, Oxford.

I. Stewart (2001). Flatterland. Macmillan, London.

2 Exploring the Fractal Universe

Arthur C. Clarke

In November 1989, when receiving the Association of Space Explorers' Special Achievement Award in Riyadh, Saudi Arabia, I had the privilege of addressing the largest gathering of astronauts and cosmonauts ever assembled at one place (more than fifty, including Apollo 11's Buzz Aldrin and Mike Collins, and the first 'space walker' Alexei Leonov ... I decided to expand their horizons by introducing them to something really large, and, with astronaut Prince Sultan bin Salman bin Abdul Azïz in the chair, delivered a lavishly illustrated lecture.

Arthur C. Clarke is the world's most prophetic and prolific writer of science fiction, much of which has become science fact. The author of well over eighty best-selling works, he is perhaps best known for originating satellite communication in 1945 and for writing the book and screenplay of *2001: A Space Odyssey*, directed by Stanley Kubrick. Arthur C. Clarke has lived in Sri Lanka since the 1950s and was knighted in 2000.

N. Lesmoir-Gordon (ed.), *The Colours of Infinity: The Beauty and Power of Fractals*, DOI 10.1007/978-1-84996-486-9_2, 2010 © The Estate of Arthur C Clarke

*... perhaps there is some structure,
if one can use that term, deep in the human mind
that resonates to the patterns in the M-set.*

Today, everybody is familiar with graphs, especially the one with Time along the horizontal axis, and the Cost of Living climbing steadily up the vertical one. The idea that any point on a plane can be expressed by two numbers, usually written x and y, now appears so obvious that it seems quite surprising that the world of mathematics had to wait until 1637 for Descartes to invent it.

We are still discovering the consequences of that apparently simple idea, and the most amazing is now just a few years old. It's called the Mandelbrot Set (from now on, the 'M-set') and you're soon going to meet it everywhere — in the design of fabrics, wallpaper, jewellery and linoleum. And, I'm afraid, it will be popping out of your TV screen in every other commercial.

The stunning beauty of the images the M-set generates has an appeal that is both emotional and universal: I have seen people almost hypnotized by the computer-produced films that explore its — literally infinite — ramifications.

Resonating to the M-set

The psychological reasons for this appeal are still a mystery, and may always remain so; perhaps there is some structure, if one can use that term, deep in the human mind that resonates to the patterns in the M-set. Carl Jung would have been surprised — and delighted — to know that thirty years after his death, the computer revolution whose beginnings he just lived to see would give new impetus to his theory of archetypes, and his belief in the existence of a 'collective unconscious'.

Many patterns in the M-set are strongly reminiscent of the abstract, curvilinear motifs of Islamic decorative art; the comma-shaped Paisley design is one example. Others resemble organic structures — tentacles, compound insect eyes, armies of sea-horses, elephant trunks ... then, abruptly, they become transformed into angular shapes like the crystals and snowflakes of the world before any life existed.

Yet perhaps the most astonishing feature of the M-set is its basic simplicity. Unlike almost everything else in modern mathematics, any schoolchild can understand how it is produced. Its generation involves nothing more advanced than addition and multiplication; there's no need for such complexities as subtraction and — heaven forbid! — division, let alone any of the more exotic beasts from the mathematical menagerie.

Another of those equations

There can be few people in the civilized world who have not encountered Einstein's famous $E = mc^2$, or who would consider it too hopelessly complicated to understand. Well, the equation that defines the M-set contains the same number of terms, indeed looks very similar. Here it is:

$$Z \rightleftharpoons z^2 + c.$$

Not very terrifying, is it? Yet the lifetime of the Universe would not be long enough to explore all its ramifications.

The zs and the c in Mandelbrot's equation are all numbers, not (as in Einstein's) physical quantities like mass and energy. They are coordinates, which specify the position of a point, and the equation controls the way in which it moves, to trace a pattern.

There's a very simple analogue familiar to everyone – those children's books with blank pages sprinkled with numbers, which when joined up in the right order reveal hidden – and often surprising – pictures. The image on a TV screen is produced by a sophisticated application of the same principle.

In theory, anyone who can add and multiply could plot out the M-set with pen or pencil on a sheet of squared paper. However, as we'll see later, there are certain practical difficulties – notably the fact that a human lifespan is seldom more than a hundred years. So the M-set is invariably computer-generated, and usually shown on a computer screen.

Take any point in space

Now, there are two ways of locating a point in space. The more common employs some kind of grid reference – West–East, North–South, or, on squared graph paper, a horizontal X-axis and a vertical Y-axis.

But there's also the system used in radar, now familiar to most people thanks to countless movies. Here the position of an object is given by (1) its distance from the origin, and (2) its direction, or compass bearing. Incidentally, this is the natural system – the one you use automatically and unconsciously when you play any ball game. Then you're concerned with distances and angles, with yourself as the origin.

So think of a computer's screen as a radar screen with a single blip on it, whose movements are going to trace out the M-set. However, before we switch

on our radar, I want to make the equation even simpler, to: $Z = z^2$. I've thrown c away, for the moment, and left only the zs. Now let me define them more precisely. Small z is the initial range of the blip – the distance at which it starts. Big Z is its final distance from the origin. Thus if a point was initially 2 units away, by obeying this equation it would promptly hop to a distance of 4.

The iteration loop

Nothing to get very excited about, but now comes the modification that makes all the difference:

$$Z \rightleftharpoons z^2$$

That double arrow is a two-way traffic sign, indicating that the numbers flow in both directions. This time, we don't stop at $Z = 4$; we make that equal to a new z – which promptly give us a second Z of 16, and so on.

In no time we've generated the series 256,65536,4294967296 … and the spot that started only 2 units from the centre is heading towards infinity in giant steps of ever-increasing magnitude.

This process of going round and round a loop is called 'iteration'. It's like a dog chasing its own tail, except that a dog doesn't get anywhere. But mathematical iteration can take us to some very strange places indeed – as we shall soon discover.

Now we're ready to turn on our radar. Most displays have range circles at 10, 20 … 100 kilometres from the centre. We will require only a single circle, at a range of 1. There's no need to specify any units, as we're dealing with pure numbers. Make them centimetres or light years, as you please.

Let's suppose that the initial position of our blip is anywhere on this circle – the bearing doesn't matter. So z is 1.

As 1 squared is still 1, so is Z. And it remains at that value because no matter how many times you square 1, it always remains exactly 1. The blip may hop round and round the circle, but it always stays on it.

Shooting for infinity

Now consider the case where the initial z is greater than 1. We've already seen how rapidly the blip shoots to infinity if z equals 2, but the same thing happens even if it's only a microscopic shade more than 1: say 1.00000000000000000001. Watch:

At the first squaring, Z becomes

   1.00000000000000000002
   then
   1.00000000000000000004
   1.00000000000000000008
   1.00000000000000000016
   1.00000000000000000032
   1.00000000000000000064

and so on for pages of printout. For all practical purposes, the value is still exactly 1. The blip hasn't moved visibly outwards or inwards; it's still on the circle at range 1.

But those zeros are slowly being whittled away, as the digits march inexorably across from the right. Quite suddenly, something appears in the third, second, first decimal place – and the numbers explode after a very few additional terms, as this example shows, reading left to right:

1.001	1.002	1.004	1.008
1.016	1.032	1.066	1.136
1.292	1.668	2.783	7.745
59.987	3598.467	12,948,970	

167,675,700,000,000

28,115,140,000,000,000,000,000,000,000

There could be a million – a billion – zeros on the right hand side, and the result would still be the same. Eventually the digit would creep up to the decimal point – and then Z would take to infinity.

The other side of infinity

Now let's look at the other case. Suppose z is a microscopic amount less than 1 – say something like 0.99999999999999999999.

As before, nothing much happens for a long time as we round the loop, except that the numbers on the far right get steadily smaller. But after a few thousand or million iterations – catastrophe! Z suddenly shrinks to nothing, dissolving in an endless string of zeros …

Check it out on your computer. It can only handle twelve digits? Well, no matter how many you had to play with, you'd get the same answer. Trust me …

The results of this 'program' can be summarized in the laws that may seem too trivial to be worth formulating. But mathematical truth is trivial, and in a few more steps these laws will take us into a universe of mind-boggling wonder and beauty.

The laws of squaring

Here are the three laws of the squaring program:

1. if the input z is exactly equal to 1, the output Z always remains 1;

2. if the input is more than 1, the output eventually becomes infinite; and

3. if the input is less than 1, the output eventually becomes zero.

That circle of radius 1 is therefore a kind of map, dividing the plane into two distinct territories. Outside it, numbers that obey the squaring law have the freedom of infinity; numbers inside it are prisoners, trapped and doomed to ultimate extinction.

At this point, someone may say: 'You've only talked about ranges – distances from the origin. To fix the blip's position, you have to give its bearing as well. What about that?'

Very true. Fortunately, in this selection process – this division of the zs into two distinct classes – bearings are

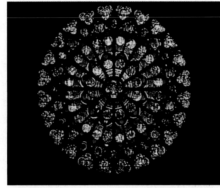

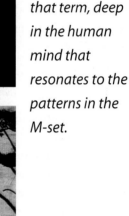

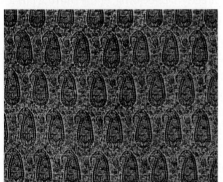

... perhaps there is some structure, if one can use that term, deep in the human mind that resonates to the patterns in the M-set.

irrelevant; the same thing happens in every direction. For this simple example – let's call it the S-set – we can ignore them.

When we come on to the more complicated case of the M-set, where the bearing is important, there's a very neat mathematical trick to take care of it. Many of you will have guessed that it uses complex or imaginary numbers (which really aren't at all complex, still less imaginary). But we don't need them for this discussion, and I promise not to mention them again.

Inside the map

The S-set lies inside a map, and its frontier is the circle enclosing it. That circle is simply a continuous line with no thickness. If you could examine it with a microscope of infinite power, it would always look exactly the same.

You could expand the S-set to the size of the universe; its boundary would still be a line of zero thickness. Yet there are no holes in it; it's an absolutely impenetrable barrier, forever separating the zs less than one from those greater than one.

Now, at last, we're ready to tackle the M-set, where these commonsense ideas are turned upside down. Fasten your seat belts.

During the 1970s, the French mathematician Benoît Mandelbrot, working at Harvard and IBM, started to investigate the equation that has made him famous, and which I will now write in dynamic form:

$$Z \rightleftharpoons z^2 + c.$$

The only difference between this and the equation we have used to describe the S-set is the term c. This – not z – is now the starting point of our mapping operation. The first time round the loop, z is put equal to zero.

The M-set and the unimaginable universe

It seems a trifling change, and no-one could have imagined the universe it would reveal. Mandelbrot himself did not obtain the first crude glimpses until the spring of 1980, when vague patterns started to emerge on computer printouts. He had begun to peer through Keats's

> Charmed magic casements, opening on the foam
> Of perilous seas, in faery lands forlorn.

As we shall learn later; that word 'foam' is surprisingly appropriate.

The new equation asks and answers the same question as the earlier one: What shape is the 'territory' mapped out when we put numbers into it? For the S-set it was a circle of radius 1. Let's see what happens when we start with this value in the M-equation.

You should be able to do it in your head – for the first few steps. But after a few dozen even a supercomputer may blow a gasket.

For starters, $z = 0$, $c = 1$. So $Z = 1$.
First loop: $Z = 1^2 + 1 = 2$.
Second loop: $Z = 2^2 + 1 = 5$.
Third loop: $Z = 5^2 + 1 = 26$.
Fourth loop: $Z = 26^2 + 1$ … and so on.

I once set my computer to work out the higher terms (about the limit of my programming ability) and it produced only two more values before it had to start approximating. Starting from the beginning we get:

1
2
5
26
677
458,330
21,006,640,000
4,412,789,000,000,000,000,000.

At that point my computer gave up, because it doesn't believe there are any numbers with more than 38 digits.

However, even the first two or three terms are quite enough to show that the M-set must have a different shape from the perfectly circular S-set. A point at distance 1 is in the S-set; indeed, it defines its boundary. A point at that same distance may be outside the boundary of the M-set.

Note that I say 'may' not 'must'. It all depends on the initial direction, or bearing, of the starting point, which we have been able to ignore hitherto because it did not affect our discussion of the (perfectly symmetrical) S-set. As it turns out, the M-set is only symmetrical about the X, or horizontal, axis.

One might have guessed that, from the nature of the equation. But no-one could possibly have intuited its real appearance: if the question had been put to me in my virginal pre-Mandelbrot days, I would probably have hazarded: 'Something like an ellipse, squashed along the Y-axis.' I might even (though I doubt it) have correctly guessed that it would be shifted towards the left, or minus, direction.

The indescribable M-set

At this point, I would like to try a thought experiment on you. The M-set being literally indescribable, here's my best attempt describe it: imagine you're looking straight down on a rather plump turtle swimming westwards. It's been crossed with a swordfish, so has a narrow spike pointing ahead of it. Its entire perimeter is festooned with bizarre marine growths – and with baby turtles of assorted sizes, which have smaller weeds growing on them …

I defy you to find a description like that in any maths textbook. And if you think you can do better when you've met the real beast, you're welcome to try.

Computers can easily make snapshots of the M- set at any magnification, and even in black and white they are fascinating. However, by a simple trick they can be coloured, and transformed into objects of amazing, even surreal, beauty.

(I suspect that the insect world might provide better analogies; there may even be a Mandelbeetle lurking in the Brazilian rain forests. Too bad, we'll never know.)

Here is the first crude approximation, shorn of details; if you like to fill its blank spaces with the medieval cartographers' favourite 'here be dragons' you will hardly be exaggerating.

First of all, note that – as I've already remarked – it's shifted to the left (or West, if you prefer) of the S-set, which of course extends from +1 to −1 along the X-axis. The M-set only gets to 0.25 on the right of the horizontal axis line itself, though above and below the axis line it bulges out to just beyond 0.4.

The 'Utter West'

On the left-hand side, the map stretches to about −1.4, and then it sprouts a peculiar spike – or antenna – which reaches out to exactly −2.0. As far as the M-set is concerned, there is nothing beyond this point; it is the edge of the Universe.

Some Mandelbrot fans call it 'the Utter West', and you might like to see what happens when you make c equal to −2. Z doesn't converge to zero – but it doesn't escape to infinity either, so the point belongs to the Set – just. But if you make c equal to −2.0000001, before you know you're passing Pluto and heading for Quasar West.

Now we come to the most important distinction between the two sets. The S-set has a nice, clean line for its boundary. The frontier of the M-set is, to say the least, fuzzy. Just how fuzzy you will begin to understand when we start to zoom into it; only then will we see the incredible flora and fauna that flourish in that disputed territory.

The boundary – if one can call it that – of the M-set is not a simple line; it is something that Euclid never imagined, and for which there is no word in ordinary language. Mandelbrot, whose command of English (and American) is awesome, has ransacked the dictionary for suggestive nouns. A few examples: foams, sponges, dusts, webs, nets, curds. He himself coined the technical name fractal, and is now putting up a spirited rearward action to stop anyone defining it too precisely.

The colours of infinity

Computers can easily make snapshots of the M-set at any magnification, and even in black and white they are fascinating. However, by a simple trick they can be coloured, and transformed into objects of amazing, even surreal, beauty.

The original equation, of course, is no more concerned with colour than is Euclid's Elements of Geometry. But if we instruct the computer to colour any given region in accordance with the number of times z goes round the loop before it decides whether or not it belongs to the M-set the results are gorgeous.

Thus the colours, though arbitrary, are not meaningless. An exact analogy is found in cartography. Think of the contour lines on a relief map, which show elevations above sea level. The spaces between them are often coloured so that the eye can more easily grasp the information conveyed. Ditto with bathymetric charts; the deeper the ocean, the darker the blue. The map-maker can make the colours anything he likes, and is guided by aesthetics as much as geography.

It's just the same here – except that these contour lines are set automatically by the speed of the calculation – I won't go into details. I have not discovered what genius first had this idea – perhaps Monsieur M. himself, but it turns them into fantastic works of art. And you should see them when they're animated …

Only in the computer age

One of the many strange thoughts that the M-set generates is this. In principle, it could have been discovered as soon as the human race learned to count. In practice, since even a low magnification image may involve billions of calculations, there was no way in which it could even be glimpsed before computers were invented! And such movies as those on the DVD with this book would have required the entire present world population to calculate night and day for years – without making a single mistake in multiplying together trillions of hundred-digit numbers.

I began by saying that the Mandelbrot Set is the most extraordinary discovery in the history of mathematics. For who could have possibly imagined that so absurdly simple an equation could have generated such literally infinite complexity, and such unearthly beauty?

The Mandelbrot Set is, as I have tried to explain, essentially a map. We've all read those stories about maps that reveal the location of hidden treasure.

Well, in this case the map IS the treasure!

The Mandelbrot Set is, as I have tried to explain, essentially a map. We've all read those stories about maps that reveal the location of hidden treasure. Well, in this case the map IS the treasure!

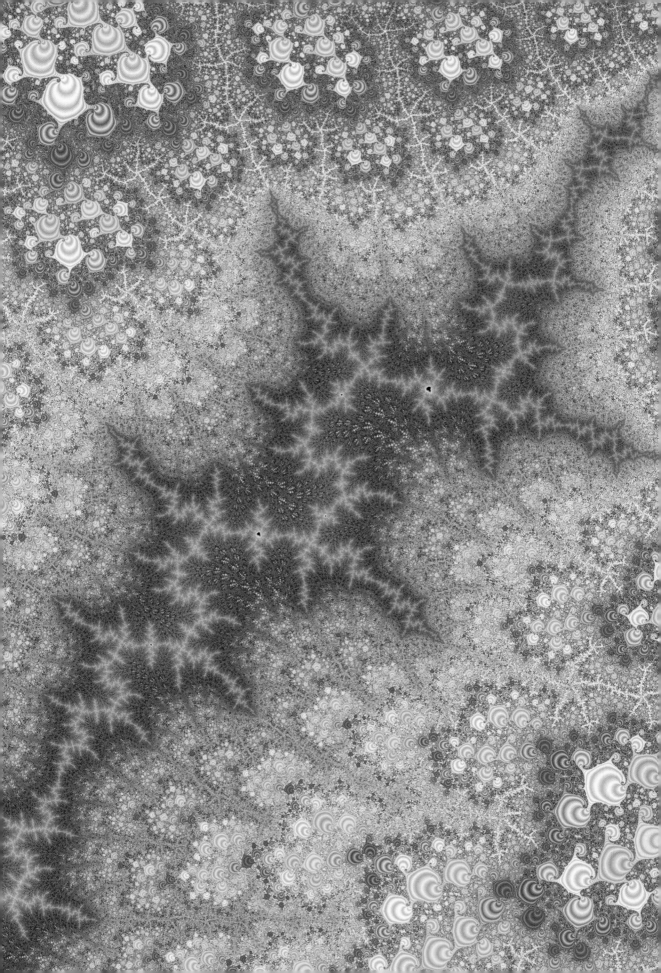

MATHEMATICAL APPENDIX

One way of appreciating where the curiously-shaped country of the M-set is located on the map of all possible (complex) numbers is to pin down its Eastern and Western frontiers, ignoring everything to the North and South.

The Western, or negative, limit is easily identified; for once, the calculation can be done mentally, without the aid of a computer! If we take the basic equation: $Z = z^2 + c$ and set the initial value of c equal to -2, the first time round the loop gives $Z = -2$. The second value is $Z = (-2)^2 - 2 = 2$. The third value is $Z = 2^2 - 2 = 2$.

And so on for ever: Z is stuck at 2! It does not shrink to zero, but neither does it go racing off to infinity. Thus the point at -2 on the X-axis, or 2 units to the left of the origin, definitely belongs to the M-set. It masks the Utter West – the very tip of the strangely ornamented spike that extends in that direction.

It's interesting to see what happens for values of c on either side of -2, and for that we certainly do need a computer. Take $c = -1.99999$. Table 2.1 shows what happens to Z as it goes round and round the loop:

Table 2.1 $c = -1.99999$ (reading the numbers left to right)

1.999970	1.999890	1.999570	1.998290	1.993174
1.972752	1.891762	1.578773	0.492534	−1.757400
1.088466	−0.815231	−1.335388	−0.216729	−1.953019
1.814292	1.291665	−0.331592	−1.890037	1.572248
0.471975	−1.777230	1.158556	−0.657737	−1.567371
0.456663	−1.791449	1.209298	−0.537588	−1.710989
0.927494	−1.139744	−0.700973	−1.508626	0.275963
−1.923834	1.701 149	0.893918	−1.200901	−0.557826
−1.688820	0.852124	−1.273875	−0.377232	−1.857686
1.451007	0.105431	−1.988874	1.955631	1.824503
1.328822	−0.234222	−1.945130	1.783540	1.181027
−0.605166	−1.633764	0.669195	−1.552168	0.409236
−1.832516	1.358123	−0.155492	−1.975812	1.903845
1.624634	0.639446	−1.591099	0.531605	−1.717386
0.949426	−1.098581			

The value then goes on oscillating, presumably forever (my computer has been round the loop only about ten thousand times) between the limits of plus and minus 2. Perhaps after ten million iterations Z might change its mind and suddenly shoot off to infinity, but it seems reasonable to assume that this value of c is definitely inside the M-set.

Table 2.2   c = −2.00001

2.00003	2.00011	2.00043	2.00171	2.00683
2.02737	2.11023	2.45306	4.01748	14.14011
197.942	39,179.2	1.5E+9		
2.4E+18	5.5E+36	3.1E+73		

The fate of the point only 0.00002 units further 'west', on the other hand, is very quickly decided, as we see in Table 2.2:

As far as an Apple Mac is concerned, the numbers in that last line are infinite, and I doubt if even a Super Cray would disagree. So −2.00001 is definitely outside the M-set.

On the Eastern, or positive side of the Set, the limit is not so easily defined.

Obviously, it is closer to the origin (0,0) than the point + 1, which gives a value shooting off to infinity after only a few times round the loop. A few minutes' work with pencil and paper shows that it is even closer than 0.5, for putting $c = \frac{1}{2}$ also gives a rapidly soaring Z. It is, in fact, at 0.25 – though this is by no means easy to prove.

When I set c = 0.25 in the program I have painfully written, the screen is flooded with a torrent of numbers, which after hundreds of iterations finally settle down to the odd value 0.4998505. I assume that this should be exactly 0.5, with the difference due to rounding-off errors. In any event, Z doesn't shoot off to infinity, so the Eastern limit of the M-set is definitely at 0.25. (On the centre line, that is; above and below, it bulges considerably further eastwards.)

It's interesting to check what happens when bracketing this value and setting c equal to 0.24999 and 0.25001. Table 2.3 gives the result of the first:

Table 2.3   c = .24999

.3124850	.3476369	.3708414	.3875133	.4181846
.4248683	.4305031	.4431468	.4463691	.4492353
.4562108	.4581183	.4598624	.4643020	.4655664
.4667420	.4698221	.4707228	.4715699	.4738342
.4745088	.4751486	.4768840	.4774084	.4779088
.4792815	.4797008	.4801028	.4812158	.4815586
.4818887	.4828091	.4830946	.4833704	.4001566
.4101153	.4353229	.4394960	.4518024	.4541154
.4614634	.4629385	.4678381	.4688625	.4723682
.4731217	.4757562	.4763340	.4783868	.4788439
.4804887	.4808594	.4822067	.4825133	.4833704

and then, after 12 more screens-full of figures …

.4968333	.4968333	.4968334	.4968334	.4968334
.4968334	.4968335	.4968335	.4968335	.4968335
.4968336	.4968336	.4968336	.4968337	.4968337
.4968337	.4968337	.4968338	.4968338	.4968338
.4968338	.4968339	.4968339	.4968339	.4968339
.4968339	.4968340	.4968340	.4968340	.4968340
.4968341	.4968341	.4968341	.4968341	.4968342

and so on forever.

$c = .24999$ is therefore definitely inside the M-set. If we increase its value very slightly, to .25001, Table 2.4 reveals a quite different result, though it takes almost as long to arrive at it.

Table 2.4 $c = .25001$

.3125150	.3476756	.3708883	.3875682	.4002191
.4305951	.4354222	.4493631	.4519372	.4600251
.4616331	.4669395	.4680425	.4718018	.4726070
.4754148	.4760292	.4101853	.4182620	.4249531
.4396025	.4432603	.4464897	.4542572	.4563596
.4582741	.4631151	.4644856	.4657569	.4690738
.4700402	.4709478	.4733673	.4740866	.4747681

and then, after some eight screens-full of figures…

.5611078	.5648520	.5690677	.5738481	.5793116
.5856120	.5929514	.6016013	.61 19342	.6244734
.6399771	.6595806	.6850566	.7193126	.7674206
.8389443	.9538376	1.159816	1.595183	2.794620
8.059914	6.5E+1	4.3E+3	1.8E+7	3.3E+14
1.07E+29	1.1E+58	1.3E+116	1.7E+232	

$c = 0.25001$ is therefore outside the M-set.

Although all these calculations involve only the X-coordinate, and ignore complex numbers by setting Y = 0, they can be very time-consuming. Tables 2.3 and 2.4 demonstrate how impossible it would have been to discover – let alone map in detail! – the Mandelbrot Set before the advent of modern computers.

3 A Geometry Able to Include Mountains and Clouds

Benoît Mandelbrot

This chapter originated in 'A Lecture on Fractals' delivered at a Nobel Conference at Gustavus Adolphus College in St Peter (Minnesota) in 1990. Mandelbrot's wide-ranging presentation and the tenor of his responses in the discussion following the lecture demonstrate the ubiquity of fractals, from nature to art and from economics to physics.

Benoît Mandelbrot was born in Poland in 1924, and moved with his family to Paris in the 1930s where one of his uncles introduced him to the Julia sets. Despite what he calls 'chaotic schooling', Mandelbrot obtained his Ph.D. in Paris in 1952. A few years later he moved to the US, and became an IBM Fellow (now Fellow Emeritus) at the Thomas B. Watson Research Center in New York. He pursued his intuitions about fractals by using the rare opportunity of massive computer power to test and prove his idiosyncratic ideas. After prestigious university appointments in a variety of subjects, he joined the Yale faculty in 1987, where he is Sterling Professor of Mathematical Science.

Mandelbrot is world-renowned for developing fractal geometry and discovering the Mandelbrot Set, named in his honour. He has written and lectured widely and has received numerous academic honours, including The Wolf Prize for Physics in 1993 and The Japan Prize for Science and Technology in 2003.

N. Lesmoir-Gordon (ed.), *The Colours of Infinity: The Beauty and Power of Fractals*, DOI 10.1007/978-1-84996-486-9_3, © Springer-Verlag London Limited 2010

In order to understand geometric shapes, I believe that one must see them

It has very often been forgotten that geometry simply must have a visual component, and I believe that in many contexts this omission has proven to be very harmful.

To begin, let me say a few words concerning the scope of fractal geometry. In 1990, I saw it as a workable geometric middle ground between the excessive geometric order of Euclid and the geometric chaos of general mathematics. It is based on a form of symmetry that had previously been underutilized, namely self-similarity, or some more general form of invariance under contraction or dilation.

Fractal geometry is conveniently viewed as a language, and it has proved its value by its uses. Its uses in art and pure mathematics, being without practical application, can be said to be poetic. Its uses in various areas of the study of materials and other areas of engineering are examples of practical prose. Its uses in physical theory, especially in conjunction with the basic equations of mathematical physics, combine poetry and high prose. Several of the problems that fractal geometry tackles involve old mysteries, some of them already known to primitive man, others mentioned in the Bible and others familiar to every landscape artist.

To elaborate, let us provide a marvellous text that Galileo wrote at the dawn of science:

Philosophy is written in this great book – I am speaking of the Universe – which is constantly offered for our contemplation, but which cannot be read until we have learned its language and have become familiar with the characters in which it is written. It is written in the language of mathematics, and its characters are triangles, circles and other geometric forms, without which it is humanly impossible to understand a single word of it; without which one wanders in vain across a dark labyrinth. (Galileo Galilei: Il Saggiatore, 1623)

We all know that mechanics and calculus, therefore all of quantitative science, were built on these characters, and we all know that these characters belong to Euclidean geometry. In addition, we all agree with Galileo that this geometry is necessary to describe the world around us, beginning with the motion of planets and the fall of stones on Earth.

A geometry of nature?

But is it sufficient? To answer, let us focus on that part of the world that we see in everyday life. Modern box-like buildings are cubes or parallelepipeds. Good-quality plasterboard is flat. Good-quality tables are flat and typically have straight or circular edges. More generally, the works of Man, as the engineer and the builder, are typically flat, round or follow the other very simple shapes of the classical schools of geometry.

By contrast, many shapes of nature – for example, those shapes of mountains, clouds, broken stones, and trees – are far too complicated for Euclidean geometry. Mountains are not cones. Clouds are not spheres. Island coastlines are not circles. Rivers don't flow straight. Therefore, we must go beyond Euclid if we want to extend science to those aspects of nature.

A geometry able to include mountains and clouds now exists. I put it together in 1975, but of course it incorporates numerous pieces that have been around for a very long time. Like everything in science, this

Fig. 3.1 left: A fractal landscape that never was (R.F.Voss).

Fig. 3.2 below: A cloud formation that never was (S. Lovejoy & B.B.Mandelbrot).

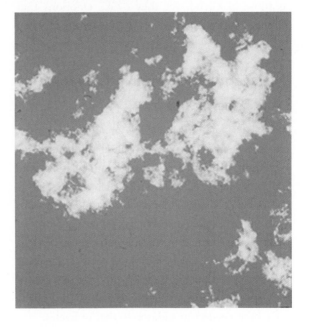

new geometry has very, very deep and long roots. Let me illustrate some of the tasks it can perform.

Figure 3.1 seems to represent a real mountain but is neither a photograph nor a painting. It is a computer forgery; it is completely based upon a mathematical formula from fractal geometry. The same is true of the forgery of a cloud that is shown in Fig. 3.2.

An amusing and important feature of these figures is that both adopt and adapt formulas that had been known in pure mathematics. Thanks to fractal geometry, diverse mathematical objects, which used to be viewed as being very far from physics, have turned out to be the proper tools for studying nature. I shall return to this in a moment.

Fractal modelling of relief was successful in an unexpected way. It is used in an immortal masterpiece of cinematography called Star Trek Two, The Wrath of Khan. Many people have seen it, but – unless prodded – few have noticed that the new planet that appears in the Genesis sequence of that film has a fractal relief. If I could show it to you, you would see that it happens to have peculiar characteristics (superhighways and square fields). They occur because of a shortcut taken by Lucasfilm in order to make it possible to compute these fractals quickly enough. But we need not dwell on flaws. Far

more interesting is the fact that the films that include fractals create a bridge between two activities that are not expected to ever meet – mathematics and physics on the one hand, and popular art on the other.

More generally, fractals have an aspect that I found very surprising at the beginning and that continues to be a source of marvel: people respond to fractals in a deeply emotional fashion. They either like them or dislike them, but in either case the emotion is completely at variance with the boredom that most people feel towards classical geometry.

Let me state that I will never say anything negative about Euclid's geometry. I love it as it was an important part of my life as a child and as a student; in fact, the main reason why I survived academically, despite a chaotic schooling, was my geometric intuition, which allowed me to cover my lack of skill as a manipulator of formulae. But we all know by experience that, apart from professional geometers, almost everybody views Euclid as being cold and dry. The fractal shapes I am showing are exactly as geometric as those of Euclid, yet they evoke emotions that geometry is not expected or supposed to evoke.

The shape of deterministic chaos

Only one new geometry

Now a few preliminary words about deterministic chaos. This topic will be touched on below, but something should be mentioned immediately. The proper geometry of deterministic chaos is the same as the proper geometry of the mountains and the clouds. Not only is fractal geometry the proper language to describe the shape of mountains and clouds, but it is also the proper language for all the geometric aspects of chaos. The fact that we need only one new geometry is really quite marvellous, because several might have been needed, in addition to that of Euclid.

I have myself devoted much effort to the study of deterministic chaos, and would like to show you now a few examples of the shapes I have encountered in this context.

Figure 3.3 is an enormously magnified fragment from a set to which my name has been attached. Here, a fragment has been magnified in a ratio equal to Avogadro's number, which is 10^{23}. Why choose this particular number? Because it's nice and very large, and such a huge magnification provides a

good opportunity for testing the quadruple-precision arithmetic on the IBM computers of a few years ago. (They passed the test. It's very amusing to be able to justify plain fun and pure science on the basis of such down-to-earth specific jobs.) If the whole Mandelbrot Set had been drawn on the same scale, the end of it would be somewhere near the star Sirius.

The shape of the black bug near the centre is very nearly the same as that of the centre of the whole Mandelbrot Set, to be discussed later when I return to this topic. Finding bugs all over is a token of geometric orderliness. On the other hand, the surrounding patterns vary from bug to bug. This is a token of variety.

The shape shown in Fig. 3.4 is a variant of the Mandelbrot Set that corresponds to a slightly different formula. This shape is reproduced here simply to comment on a totally amazing and extraordinarily satisfying aspect of fractal geometry. Fractals are perceived by many people as being beautiful, but were initially developed for the purpose of science, for the purpose of understanding how the world is put

Fig. 3.3 above: A very small fragment of the Mandelbrot Set (R.F.Voss).

Fig. 3.5 above: Cauliflower *Romanesco*.

Fig. 3.4 below: A small fragment of a modified Mandelbrot Set (B.B.Mandelbrot).

together – both statically (in terms of mountains) and dynamically (in terms of chaos, strange attractors etc.).

In other words, the shapes shown in Figs. 3.1 to 3.4 were not intended to be beautiful. So why is it that they are perceived as beautiful? The fact that they are must tell us about something regarding our system of visual perception.

I started with these four figures because their structure is so rich, but I went overboard. The richness of their structure means that these figures cannot be used to explain the main feature of all fractals. The underlying basic principle shows far more clearly on Fig. 3.5, which – for a change – reproduces a real photograph of a real object. You may recognize the Romanesco variety of cauliflower. Each bud looks absolutely like the whole head, and in turn, each bud subdivides into smaller buds, and so on. I am told that the same structure repeats over five levels of separation that you can see with the naked eye, and then through many more levels that you can only see with a magnifying glass or microscope.

Scientists' first reaction to such shapes was to focus on the spirals formed by the buds. This interest led to

extensive knowledge about the relation between the golden mean (and the Fibonacci series), and the way plants spiral. But to me what is more important is the hierarchical structure of buds because it embodies the essential idea behind fractals.

What is a fractal?

Before we go on to tackle what a fractal is, let us ponder what a fractal is not. Zoom on to a geometric shape and examine it in increasing detail. That is, take smaller and smaller portions near a point P, and allow every one to be dilated, that is, enlarged to some prescribed overall size.

If our shape belongs to standard geometry, it is well known that the enlargements become increasingly smooth. That is, one expects a curve to be 'attracted', under dilations, towards a straight line (thus defining the tangent at the point P). The term 'attractor' is borrowed from dynamics and probability theory. One also expects a curve to be attracted under dilation to a plane (thus defining the tangent plane at the point P).

An exception to this rule is when P is a double point of a curve; the curve near P is then attracted to two intersecting lines and has two tangents, but double points are few and far between in standard curves. In general, one can say that nearly every standard shape's local structure converges under dilation to one of the small number of 'universal attractors'. The grandiose term universal is borrowed from recent physics.

Yet the shapes I have been showing fail to be locally linear. In fact, they deserve to be called 'geometrically chaotic' until proven otherwise. In an isolated neighbourhood of the great City of Science, a kind of geometric chaos was discovered in the fifty years from 1875. Then, while trying to escape their concern about nature, mathematicians became aware of the fact that a geometric shape's roughness need not vanish as the examination becomes more searching. It is conceivable that it should either remain constant, or endlessly vary up and down.

The hold of standard geometry was so powerful, however, that the resulting shapes were not recognized as models of nature. Quite to the contrary, their discoverer proudly labelled them 'monstrous' and 'pathological'. After discovering these sets, mathematics proceeded to increasingly greater generality.

Like a sailor, science must constantly navigate between two dangers: the lack of and excess of generality. Between the extremes of the excessive geometric order of Euclid, and of the geometric chaos of the most general mathematics, can there be a middle ground? To provide one is the ambition of fractal geometry.

The essential nature of fractals

The reason why fractals are far more special than the most general shapes of mathematics, is because they are characterized by so-called 'symmetries', which are invariances under dilations and/or contractions. Broadly speaking, mathematical and natural fractals are shapes whose roughness and fragmentation neither tend to vanish, nor fluctuate up and down, but remain essentially unchanged as one continually zooms in. Hence, the structure of every piece holds the key to the whole structure.

The preceding statement is made precise and illustrated by Fig. 3.6, which represents a shape that is enormously more simple than those shown previously. As a joke, I called it the 'Sierpinski gasket', and the joke has stuck.

The four small diagrams show the 'initiator' of the construction, which is a triangle, then its first three stages, while the large diagram shows an advanced stage. The basic step of the construction is to divide a given (black) triangle into four sub-triangles, and then erase (whiten) the middle fourth. This step is first performed with a wholly black filled-in triangle of side 1, then with three remaining black triangles of side $^1/_2$. This process continues, following a pattern called recursive deletion, which is very widely used to construct fractals. Related patterns are recursive substitution and recursive addition (which we shall encounter) and recursive multiplication (which is fundamental but beyond the scope of this talk).

Now, take the gasket and perform an isotropic linear reduction whose ratio is the same in all directions – namely $^1/_2$ – and whose fixed point is any of the three apexes of the initiator triangle. This transformation is called a similarity. More precisely, it is homothety or linear self-similarity. By examining the large advanced stage picture, it is obvious that each of the three reduced gaskets is simply superposed on one-third of the overall shape. For this reason, the fractal gasket is said to have three properties of self-similarity.

The essence of self-similarity

Precise terminology is necessary here because one can also understand 'similar' as a loose everyday synonym of 'analogous'. In the early days of fractal geometry, the resulting terminological ambiguity was acceptable to physicists, because early detailed studies did indeed concentrate on linearly self-similar shapes. However, later developments have extended to self-affine shapes, in which the reductions are still linear, but the reduction ratios in different directions are different. For example, in order to go from a large to a small piece of fractal relief, one must contract the horizontal and vertical coordinates in different ratios. Hence, a fractal relief is called linearly self-affine.

When the Sierpinski gasket is constructed by deleting middle triangles, as in Fig. 3.6, its self-similarity

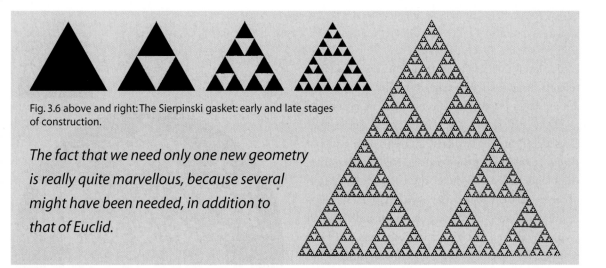

Fig. 3.6 above and right: The Sierpinski gasket: early and late stages of construction.

The fact that we need only one new geometry is really quite marvellous, because several might have been needed, in addition to that of Euclid.

seems, so to speak, to be 'static' and 'after-the-fact'. But this is a completely misleading impression. Its prevalence and its being viewed as a flaw are continual sources of surprise. In fact, the same symmetries can be reinterpreted 'dynamically' and suffice to generate the gasket. The device, which is called the 'chaos game', is a stochastic, or randomly determined interpretation of a scheme made by Hutchinson. Start with an 'initiator', that is, an arbitrary bounded set, for example a P_0. Denote the three similarities of the gasket by S_0, S_1, and S_2, and denote by $k(m)$ a random sequence of the digits 0, 1 and 2. Then define an 'orbit', as made of the points

$P_1 = S_{k(1)} (P_0)$, $P_2 = S_{k(2)} (P_1)$ and more generally $P_j = S_{k(j)} (P_{j-1})$. One finds that this orbit is 'attracted' to the gasket, and that after a few stages it describes its shape very well.

In 1964, when I first used the word 'self-similarity', I thought it was a neologism. In fact at least one writer had used it before. But the idea itself is perfectly obvious and must be very old. The reason the word was needed is that the shapes to which it refers had no importance until my work. For example, Sierpinski had defined his shape for some purpose that has long been forgotten – because it was not very important.

Why did self-similarity become important? Because Figs. 3.1 to 3.5 are self-similar, not – to be sure – in an exact, but in a slightly loose meaning of the word. Why fractal geometry has become such a large sub-

ject, and why I spent so much time in my efforts to build it as a discipline, is driven by the empirical discoveries (each established by a separate investigation) that the relief of planet Earth is self-similar, and that the same is true of many other shapes around us. Figures 3.1 to 3.5 suffice to show that the impression that self-similarity is a barren and not very fruitful idea would be an altogether wrong impression.

Granted what has just been asserted, why did the gasket become important? It does not represent anything of interest; in fact, it is so relentlessly monotonous that it could be seen as being as simple as Euclid. You can know nearly everything about it in just a few days of study. The same holds for another widely known shape, called the snowflake curve or Von Koch Island, for a set Cantor Dust, and for a few other long-known structures of the same ilk. The reason why they are important is because you must begin the study of fractal geometry with the Sierpinski gasket and its type, but keep in mind that the real fun begins beyond them.

The new Peano curve

The fun begins after one has added an element of unpredictability, due to either randomness (as in Figs. 3.1, 3.2 and 3.5) or non-linearity (as in Figs. 3.3 and 3.5). Non-linearity is the key word of the new meaning of chaos, namely of deterministic chaos, and randomness is the key to chaos in the old sense of the word. The two are very intimately linked.

But let us not rush away from linearly self-similar fractals, because in some cases a suitable graphic rendering suffices to break their relentless monotony.

Figure 3.7 shows my variant of a curve that Giuseppe Peano constructed in 1890. The point of Peano curves is that they manage to fill a portion of the plane, hence contradict the basis of the notion of curves. Mathematicians have written pages and pages to praise the freedom of imagination that allows man to invent shapes that are completely removed from reality. The Peano curve was specifically designed to be a counterexample to a natural belief that used to be universal: that curves and surfaces do not mix. It was designed for the purpose of separating mathematics and physics into two completely independent investigations. Unfortunately, it was quite successful in that respect, at least for a century.

To obtain my new Peano curve, you replace an initial straight segment by the complicated zigzag (top left). Then (top middle) each zig and zag is replaced by smaller versions of the zigzag on the top left. The same pattern (called recursive substitution) is then repeated without end. In the top-right diagram, it is easy to believe that the boundary between black and white will end up filling a snowflake curve. I call it a 'snowflake sweep'. The bottom of Fig. 3.7 reproduces the same curve but will replace every segment by an arc of a circle.

This fancy computer rendering was great fun but had a very practical goal. It was carefully thought through to force everybody to see all kinds of branching systems of arteries and veins, or of rivers, or of flames or whatever else you prefer. But those very realistic things were not seen until my work, if only because mathematicians spurned their ability to see. Partly as a result, mathematics and physics did indeed move in very different directions.

Figure 3.8 combines a sequence of completely artificial, random landscapes. Each part of this picture consists of enlarging a small black rectangle in the preceding picture and then filling in additional detail. This procedure is called recursive addition. Each

Fig. 3.7 above: Mandelbrot's Peano curve (B.B.Mandelbrot).

landscape differs from the preceding one by being more detailed, yet at the same time the successive enlargements are comparable. They might have been different parts of the same coastline examined on the same scale, but in fact they are neighbourhoods of one single point examined at very different scales. Clearly, these successive enlargements of a coastline completely fail to converge to a limit tangent!

How to measure roughness

At this point, let me recall a story about the great difficulties the ancient Greeks used to experience in formulating the idea of 'size'. Navigators knew that Sardinia took longer to circumnavigate than Sicily. On the other hand, there was evidence that Sardinia's fields are smaller than Sicily's. So which was the bigger island? Greeks sailors seem to have held the belief that Sardinia was bigger because its coastline was longer.

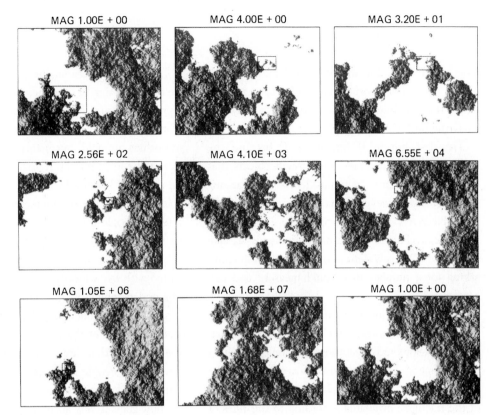

MAG 1.00E + 00 MAG 4.00E + 00 MAG 3.20E + 01

MAG 2.56E + 02 MAG 4.10E + 03 MAG 6.55E + 04

MAG 1.05E + 06 MAG 1.68E + 07 MAG 1.00E + 00

Fig. 3.8 right: Zoom onto a fractal landscape that never was (R.F.Voss).

Fig. 3.9 above: A fractal coastline that never was (B.B.Mandelbrot).

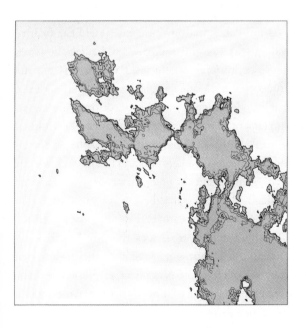

But let us examine Fig. 3.9, and ponder the notion of coastline length. When the ship used to circumnavigate is large, the captain will report a rather small length. A much smaller ship would come closer to the shore and navigate along a longer curve. A man walking along the coastline will measure an even longer length. So what about the 'real length of the coast of Sardinia'? The question seems both elementary and silly, but it turns out to have an unexpected answer. The answer is, 'it depends'. The length of a coastline depends on whether you circumnavigate it in a large or a small ship, or walk along it, or use a mouse or some other instrument to measure the coastline.

This makes us appreciate the extraordinary power of the mental structure that schools have imposed by restricting their teaching of geometry to Euclid. Many people thought they never understood geometry, yet they learned enough to expect every curve to have a length. For the curves in which I am interested, this turns out to have been the wrong thing to remember from school. Once again, the theoretical length is infinite, and the practical length depends on the method of measurement. Its increase is faster where the coastline is rough, making it necessary to study the notion of roughness.

This last notion is fundamental, because the world we live in includes many rough objects that can cause great harm. Man must learn to live among those objects. However, the task of measuring roughness objectively has turned out to be extraordinarily difficult. People whose work demands it, like metallurgists, merely went to their friends in statistics asking for a number they could measure and call roughness.

But the following experiment reveals a serious problem. Take samples of steel that the US National Bureau of Standards guarantees to come from one block of metal as as homogeneous as man can make it. Break the steel samples and measure the roughness of the fractures, evaluated according to the rules of statistics. You will find that the values you get are in complete disagreement.

Fractal dimension is the answer

On the other hand, I argue that roughness happens to be measured consistently by a quantity called fractal dimension, which happens in general to be a fraction, and which one can measure very accurately. Studying many samples from the same block of metal, we found the same dimension for every sample.

So what about the 'real length of the coast of Sardinia'? The question seems both elementary and silly, but it turns out to have an unexpected answer...it depends on whether you circumnavigate it in a ..ship...or walk along it, or use a mouse or some other instrument to measure the coastline.

The idea is that fractal dimension is a proper measure for the notion of roughness just as temperature is a proper measure for the notion of hotness. Man must have known forever that some things are hot and others are cold, but before physics could move on to a theory of matter, it was necessary to describe the degree of hotness by one number. This was possible only when the thermometer was invented, and different people using the same thermometer could get the same value of hotness for the same object.

Similarly and most fortunately, fractal geometry started with a few ideas about how to express roughness and complexity by a number. Some of these ideas add up a bunch of related but distinct tools (one can think of them as being different types of screwdrivers) that are collectively called 'fractal dimensions'. People who work with fractal geometry quickly develop an intuition of fractal dimension and can now guess it very accurately for simple shapes.

The reason we use the term 'dimension' is that it can also be applied to points, intervals, full squares and full cubes, and in those cases yield the familiar values of 0, 1, 2 and 3. Applied to fractals, however, these definitions usually yield values that are not integers. The loose idea of 'roughness' has turned out to demand a number of distinct numerical implementations, hence the multiplicity of distinct 'fractal dimensions' has proven valuable. A dimension delineated by Hausdorff and Besicovitch was the first example, but for practical needs it is either too difficult or too specialized.

The simplest variant is the similarity dimension D_s, which applies to shapes that are linearly self-similar. As I have already stated, this means that they are made up of N replicas of the whole, each replica being reduced linearly in the same ratio r. Then one defines

$$D_s = \frac{\log N}{\log(^1/_r)}$$

For a point, an interval, a square and a full cube, one has D_s = 0, 1, 2 and 3, respectively. As announced, these are the familiar values of the 'ordinary' dimensions. But the Sierpinski gasket adds something very new: one has N = 3 and $r = 1/_2$, hence

$$D = \log 3 / \log 2 \sim 1.5849...$$

Another simple fractal dimension is the mass dimension. Take a distribution of mass of uniform density on the line, in the place or in space. Then choose a sphere of radius R whose centre lies in our set. The mass in such a sphere takes the form MR = FR^D, where D is the 'ordinary' dimension and F is a numerical constant. The idea of uniform density extends to fractals, and in many cases an exponent D can be defined; it is called the mass dimension and is often equal to the similarity dimension. Unfortunately, we must move away from dimension.

How to grow a tree

The next subject I wish to tackle is the increasingly valuable role of fractal geometry as a tool in the discovery and study of previously unknown aspects of nature. Nothing illustrates this role better than a form of random growth that generates the Fractal Diffusion Limited Aggregates (DLA) or Witten-Sander aggregates. A DLA cluster lurks in the centre of Fig. 3.10. It is a tree-like shape of baffling complexity that one can use to model how ash forms, how water seeps through rock, how cracks spread in a solid and how lightning discharges.

To see how the growth proceeds, take a very large chess board and place a queen that is not allowed to move in the central square. Pawns are allowed to move in any of the four directions on the board. They are released from a random starting point at the edge of the board, and are instructed to perform a random or drunkard's walk. Each step can take one of four directions chosen with equal probabilities. When a pawn reaches a square next to that of the original queen, it transforms itself into a new queen and cannot move any further. Eventually, one has a branched, spidery collection of queens.

Quite unexpectedly, massive computer simulations show that DLA clusters are fractal. They are nearly self-similar, that is, small portions are very much like reduced versions of large portions. But deviations from randomized linear self-similarity are obvious and pose interesting challenges.

One reason for the importance of DLA is that it concerns the interface between the smooth and the fractal. A premise of fractal geometry is that much in the world is fractal. Nevertheless, science is expected to be cumulative, the new being added to the old, without chasing it away. Therefore, new wisdoms must not deny the old wisdom that the world is made of smooth shapes and involves smooth variation and differential equations.

Fig. 3.10 above: A cluster of diffusion limited aggregation, surrounded by its equipotential curves (C.J.G.Evertsz and B.B.Mandelbrot).

What DLA shows is that the old and new wisdoms are compatible only if one abandons the old philosophical expectation that everything in the world will eventually prove to be smooth or of smooth variation.

To show how smooth variation can produce rugged behaviour, the original construction must first be rephrased in terms of the theory of electrostatic potential. The description that follows is necessarily a little schematic. Grow DLA in the big box connected to a positive potential (to be taken as unity) and connect the cluster itself to the potential 0. Then the value of the potential elsewhere in the box is best described by equipotential curves, for example, the curves along which the potential takes the increasing values .01, .02, … .99.

Figure 3.10 shows that all these curves are smooth and that they provide a progressive transition between the box and the boundary of the cluster. Analytic calculation is out of the question, but 'physical common sense' can be combined with numerical calculation. In effect, the object's boundary includes many needles, and each has a high probability of getting hit by lightning. This is manifested by the fact that equipotential lines crowd together near the tips of a DLA cluster. More generally, returning to the random pawns that build up a DLA cluster, the position where the pawn lands is obtained from the shapes of the electrostatic equipotentials.

Now we come to the next logical step, which implies that DLA has brought an intellectual innovation of the highest order. For nearly 200 years, the study of potentials has limited itself to fixed boundaries. But in the simple random walk that creates DLA, a 'hit' in the above terminology can be interpreted as provoking a displacement of the boundary. Thus, the massive numerical experiments about DLA teach us that when one allows boundaries to move in response to the potential, the boundaries become fractal.

This shows without any trace of doubt that one can create rough fractals from the smoothness that characterizes equipotential lines, but this knowledge remains imperfect. We all thirst for new mathematics and physics. Nevertheless, it is worth noting how fractal geometry has led to an altogether new problem, outlined the broad solution and set many scientists to work.

The Julia Set

Our next move returns from randomness to deterministic chaos, and replaces objects in real physical space by imaginary objects. What will remain unchanged is that we shall deal with spiky sets surrounded by smooth equipotential lines.

The first notion here is that of the Julia Set of quadratic iteration. Pick a point c of coordinates u and v, and call it a 'parameter'. Next, in a different plane, a point P_0 of the coordinates x_0 and y_0. Then form $x_1 = x_0^2 - y_0^2 + u$ and $y_1 = 2x_0y_0 + v$. These formulas may seem a bit artificial, but in order to satisfy the reader who is scared of complex numbers, they simplify if the point c of coordinates x and y is represented by the complex number $z = x + iy$. (Complex numbers add and multiply like ordinary numbers, except that i^2 must always be replaced by -1.) In terms of the complex numbers $c = u + iv$ and $z = x + iy$, the preceding rule simplifies to $z_1 = z_0^2 + c$ and (more generally) $z_{k+1} = z_k^2 + c$. Even the reader who is scared of complex numbers is able to understand the expressions in terms of x_k and y_k.

When the orbit P_k fails to escape to infinity, the initial P_0 is said to belong to the 'filled-in Julia Set'. An example is shown in Fig. 3.11. If you start outside of the black shape, you go to infinity. If you start inside, you fail to iterate to infinity.

The boundary between black and white is called a 'Julia curve'. It is approximately self-similar. Each chunk is not quite identical to a bigger chunk, because of non-linear deformation. Yet, it is astonishing that iteration should create any form of self-similarity, quite spontaneously.

As in the investigation of fractal mountains, the computer was essential to the study of iteration. The bulk of fractal geometry is concerned with shapes of great apparent complication and they could never be drawn by hand. More precisely, this picture might have been computed by a hundred different people working for years, but nobody would have started such an enormous calculation without first feeling that it was worth performing.

Not only did I have access to a computer in 1979, but I was familiar with its power. The fact that no one knew what was going to emerge was enough to make these calculations worth trying. A fishing expedition led to a primitive form of Fig. 3.12. The Julia Sets of the map $z^2 + c$ can take all kinds of shapes, and a small change in C can change the Julia Set very greatly. I set out to classify all the possible shapes (for reasons that are too lengthy to discuss) and came up with a new shape. That it has been called the Mandelbrot Set is of course a great honour. Figure 3.3 above was a tiny portion of Fig. 3.12.

Fig. 3.11 above: Quadratic Julia Set for the map $z \rightleftharpoons z^2 + C$.

Constructing the Mandelbrot Set

Here is how the Mandelbrot Set is constructed. Take a starting point C_0 in the plane of coordinates u_0 and v_0. From the coordinate of C_0, form a second point C_1 of coordinates $u_1 = u_0^2 - v_0^2 + u_0$

and $v_1 = 2u_0 u_0 + v_0$.

Next, form the point C_2 of coordinates

$$u^2 = u_1^2 - v_1^2 + u_0$$

and

$$v_2 = 2u_1 u_1 + v_0.$$

More generally, the coordinates u_k and v_k of C_k are obtained from u_{k-1} and v_{k-1} by the so-called 'iterative formulas'

$$u_k = u_{k^2 - 1} - v_{k^2 - 1} + u_0$$

and

$$v_k = 2u_{k-1} v_{k-1} + v_0.$$

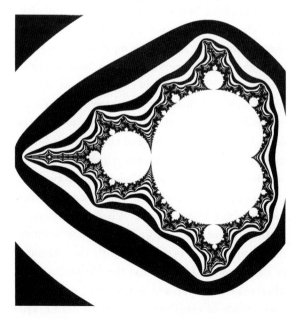

Fig. 3.12 below: The Mandelbrot Set, surrounded by its equipotential curves.

When C_0 is represented by $z_0 = u_0 + iv_0$, the above formulas simplify to $z_1 = z_0^2 + z_0$ and $z_k = z_{k-1}^2 + z_0$. The points C_k are said to form the orbit of C_0, and the set M is defined as follows: If the orbit C_k fails to go to infinity, one says that C_0 is contained within the set M. If the orbit C_k does go to infinity, one says that the point C_0 is outside M.

This algorithm concerns the following very sober problem of deterministic dynamics. When C_0 is in the interior of M, quadratic dynamics yields an orbit that is perfectly orderly, in the sense that it is asymptotically periodic. When C_0 is outside M, to the contrary, the behaviour of the orbit is deterministic, but almost unpredictably, chaotic. Quadratic dynamics was singled out for detailed study because in this case the criterion separating orderly from chaotic behaviour is as clean as can be, as seen above. The boundary between the two possibilities turns out to be messy beyond any expectation.

Zooming towards a portion of the boundary of the Mandelbrot Set, you see two distinct phenomena. The part is simply a repetition of something already seen. This element of repetition is essential to beauty. But beauty also requires an element of change, and this is also very clearly present. As you come closer and closer, what you see becomes more and more complicated. The overall shape is the same, but the hair structure becomes more and more intense. This feature is not something we put in on purpose. In so far as the mathematics is concerned, it is not invented, but discovered: we see something that has been there forever. What we discover is that the mathematics of z squared plus C is astonishingly complicated, by contrast with the simplicity of the formula. We find that the Mandelbrot Set, when examined more and more closely, exhibits the co-existence of something that repeats itself relentlessly, something that exhibits a variety that boggles the imagination. I first saw the Mandelbrot Set on a black and white screen of very low graphic quality,

and the picture looked dirty. But zooming in on what seemed like dirt revealed an extraordinary little copy of the whole.

In Fig. 3.12, the Mandelbrot Set is the white 'bug' in the middle. It is very rough-edged, but is surrounded by a collection of zebra stripes whose edges become increasingly smooth as one goes away from M. These zebra-stripe edges happen to be Laplacian equipotential curves. They are just like those in Fig. 3.10 but are far easier to obtain.

Fractal art and the mathematician

To the layman, fractal art tends to seem simply magical, but no mathematician can fail to try to understand its structure and meaning. A remarkable aspect of recent events is that the mathematics triggered by the Mandelbrot Set could have passed as 'pure' if only its visual origin could have been hidden. To many mathematicians, the newly opened possibility of playing with pictures interactively has revealed a new mine of purely mathematical questions and conjectures, of isolated problems and whole theories. To take an example, examination of the Mandelbrot Set led me in 1980 to many conjectures that were simple to state, but then proved very hard to crack. (The main one remains unsolved.) To mathematicians, their being difficult and slow to develop does not make them any less fascinating, because a host of intrinsically interesting side-results have been obtained in their study.

Herein lies a tale. Pure mathematics is certainly one of the remarkable activities of man; it certainly is different in spirit from the art of creating pictures by numerical manipulation, and it has indeed proven that it can thrive in splendid isolation – at least over some brief periods. Nevertheless, the interaction between art, mathematics and fractals confirms what is suggested by almost all earlier experiences. Over the long haul, mathematics gains by not attempting to destroy the organic unity that appears to exist between seemingly disparate but equally worthy activities of man, the abstract and the intuitive.

Of course, the black and white figures in this chapter are not beautiful colour fractal pictures. As in the case of the mountains, the quality of the colour rendering shows the skills of the programmers, but the structure itself is independent of the colour rendering. What is important is that the structure is too complicated to be understood unless the colour rendering is sufficiently rich. In fact, the set has such an enormous amount of structure that we cannot see it in one single colour rendering. Different renderings emphasize very different aspects of it. Again, this structure was not invented for the purpose of doing something beautiful, but purely for the purpose of exploring the advanced theory of z squared plus C.

Simplicity generates marvellous complexity

Let me now bring together the separate strings of my chapter. How did fractals come to play their role of 'extracting order out of chaos'? The key resides in a very surprising discovery that I made thanks to computer graphics.

The algorithms that generate fractals are typically so extraordinarily short as to look positively dumb. This means they must be called 'simple'. Their fractal outputs, on the contrary, often appear to involve structures of great richness. A priori, one would expect the construction of complex shapes to necessitate complex rules, but surprisingly, it is not so.

What is the special feature that makes fractal geometry perform in such an unusual manner? The answer is very simple. The algorithms are recursive, and the computer code written to represent them involves 'loops'. That is, the basic instructions are simple, and their effects can be followed easily.

Let these simple instructions be followed repeatedly. Unless one deals with the simple old fractals (the Cantor Set and Sierpinski gasket), the process of iteration effectively builds up an increasingly complicated transform, whose effects the mind can follow less and less easily. Eventually, one reaches something that is qualitatively different from the original building block. One can say that the situation is a fulfilment of what in general is nothing but a dream: the hope of describing and explaining chaotic nature as the cumulation of many simple steps.

Many fractals have been accepted as works in a new form of art. Some are representational, while others are totally unreal and abstract, yet all strike almost everyone in forceful, almost sensual, fashion. The artist, the child and the 'man in the street' can never see enough as they never expect to get anything of this sort from mathematics.

Nor did mathematicians expect their subject to interact with art in this way. Eugene Wigner has written about 'the unreasonable effectiveness of mathematics in the natural sciences'. To this line, I have been privileged to add another parallel statement, concerning 'the unreasonable effectiveness of mathematics as creator of shapes that Man can marvel about, and enjoy'.

After Benoît Mandelbrot had delivered this paper, he answered some questions:

Chairman: First of all, are there any responses from the panel?

Q1: Benoît, if I could ask, speaking of poetry and prose, this is a rather flippant question, but is it more like music or like noise?

Mandelbrot: For me, music is a form of poetry, and I forgot to say so simply because I felt it was obvious. Analogies can become very dangerous if pursued too far, but I'm glad that you have been taken by the game.

Q2: Benoît has given us a lot of very nice insights, I think, into this kind of geometry, but I would like to express a little, maybe different point of view, which emphasizes something else. I think Benoît spoke at one point, something to the effect that the physical properties reduced to the geometric properties. And I think, somehow the geometry is very static, and to me the static should best be seen with deeper understanding as flowing from the dynamics. Therefore, I would put a dynamical perspective on the understanding of physics, above that of a geometrical perspective. In the dynamics, the physical process itself, the equations, which are time-dependent, from those one can derive some of these fractal geometric pictures with a deeper understanding than just looking at the pure fractal geometry in its own right. So, for me, there's a little more primary emphasis on the deeper physics coming from the dynamics rather than the geometry.

Mandelbrot: I see absolutely no conflict between our viewpoints. To study the dynamics of Julia Sets, you must study the statics of the Mandelbrot Set. In many cases, for example, the shape of the mountains, everyone knows well, is static. If so, the next step would be to understand the processes that create the mountains. This task is far from having been completed, but James Bardeen has constructed in successive fractal pictures that attempt to make use of what is known of the dynamics in order to represent the statics. Since very often the geometry of statics is fractal, and the geometry of dynamics is also fractal, fractals do not lose either way.

Q3: I would like to say that I'm completely in agreement with what Q2 just said. I believe that one of the main points is to relate dynamics to chaos and to fractals. In fact, let me give two examples where I think some additional dynamics would be very nice. When we speak about adding some

noise, from where is this noise coming? And when we speak about boundary conditions, from where are the boundary conditions coming? Essentially, boundary conditions are an empirical concept. In hydrodynamics or microscopic physics, you can speak about boundaries. If you speak about dynamics, there are no boundaries. Boundaries are part of the dynamical problem. Therefore, in a sense, I think that your presentation, which was very beautiful, of course, is more a kind of phenomenology which has to be, I would say, made a little deeper by making some relation with dynamical concepts.

Mandelbrot: Two of the figures illustrated a fractal aggregate. As it grows, its boundary is continually changed by the dynamics of the generating process. Thus, I agree with what you say. This dynamics consists in little particles aggregating together, but eventually leads to an extraordinary structure. The open mystery is why this structure is fractal.

Chairman: In the past, large mathematical models were used to centralize decisions – for example, in economics – for traditional models have not worked. What does the new science bring to prediction, control and, ultimately, to social responsibility?

Mandelbrot: Your question is complicated. I prefer not to answer the last part.

But I have been greatly interested in economics. In view of your comments, I must emphasize that existing economic thought strikingly fail to predict anything about those aspects of the economy on which tests are possible, because data are available in large quantity. For example, many people attempt to explain or predict the stock market, but they all fail. My approach to finance in the early 1960s was very different. It was phenomenological, absolutely, deliberately, and even arrogantly. My goal was to generate wiggles that people active in the stock market would not be able to distinguish from the wiggles they see in newspapers. This goal was both modest and demanding; I succeeded with the help of a very simple, purely random process. Economists challenged me to explain my statistical statics from their dynamics. Disappointingly, their dynamic is not up to the task.

Economics and other more complicated areas borrow a great deal from physics. What they borrow is mostly made of fully developed concepts and theories, such as the concept of equilibrium and the theory of displacement of equilibrium in perfect gases. Next, they try to develop these concepts and themes in rigorous fashion in an economics context. Much less effort is devoted to testing whether economic phenomena really fall into the domain in which those standard physical arguments can conceivably apply.

For example, take continuity. Everyone in economics seemed to assume that prices were a continuous function of time. To the contrary, all the evidence shows that one comes much closer to reality by assuming prices to be a discontinuous function of time. Incidentally, this discontinuity is not that of quantum physics.

To summarize: I was active in economics both in the early 1960s and again more recently. The reason why my effort in this area has been arrogantly

phenomenological is because the more ambitious dynamical study of these things has been an abject failure.

Q4: I was interested to see that that question was put in the past tense about complicated economic models. It continues to be true that a fantastic amount of money and effort is put into enormously complex, many variable, mostly linear, economic forecasting models. You can read the predictions from these every year in The Wall Street Journal. Models that attempt to link tens of thousands of variables and relationships – home mortgage interest rates, the ratio of the dollar and the yen, the demand for Sierpinski gaskets – anything you can imagine is built into these models, and the results are often announced to two or three digits of precision. And then, of course, next year, they have to be artificially amended with tens of thousands of ad hoc changes. I think we're only beginning to see an appreciation by some economists of some of the work you've already started to describe, and that you'll hear described as this conference goes on. An appreciation of what can be done with a greater recognition of the essential non-linearity of enormous complex systems like economics.

Chairman: I have one more question from the audience: When doing mathematical research, do you discover or invent?

Mandelbrot: I certainly feel that I discover. The assertion that eventually became the four-colour theorem was discovered long ago … by an amateur. It was not some new thing to be invented, but an existing fact to be discovered. It was there.

The same was true when I sat in front of a terminal, next to this extraordinarily gifted young assistant, to investigate the set that became known as the Mandelbrot Set. It was never our feeling that we were inventing anything. This thing was there. My whole thrust was to discover more about its complication. Its complication was the key to the dynamics of quadratic iteration, which is a dynamical system with particularly simple equations. We tried to discover the so-called static geometry of one set, in order to understand the dynamics of another set.

Let me also mention the work on multifractals that I did in the 1960s and published in 1974. In this instance, the process of discovery occurred on two levels. First of all, I discovered new facts about random singular measures. The key was a mathematical theorem that I had learned as a young man, but had always felt would never be used in physics. Hence, it is the study of multifractals that made me discover the real meaning of that theorem. Until then, its statement was so abstract that I could not see it and appreciate what it had always meant.

Proofs are very often a very different matter. Some are so contrived that they definitely look and feel invented, but the best proofs also have both the look and the feel of discovery.

Further reading

1. The Fractal Geometry of Nature by B.B. Mandelbrot (W.H. Freeman, 1982) was the first comprehensive book on the subject, and remains a basic reference book. Innumerable other books have appeared since. An up-to date list is found on the website www.math.yale.edu/mandelbrot

2. The basic how-to book is The Science of Fractal Images, eds. H.-O. Peitgen and D. Saupe (Springer, 1988).

3. The best-known book on iteration is, deservedly, The Beauty of Fractals by H.-O. Peitgen and P.H. Richter (Springer, 1986).

4. For other aspects of the mathematics, see Fractals: Mathematical Foundations and Applications by K.J. Falconer (Wiley, 1990) and Fractal Geometry and its Applications: a Jubilee of B. Mandelbrot ed. M. Lapidus (2004)

On the concrete uses of fractals, three references are convenient, because they are special volumes of widely available periodicals:

5. Proceedings of the Royal Society of London, Volume A423 (8 May 1989), which was also reprinted as Fractals in the Natural Sciences, ed. M. Fleischmann et al. (Princeton University Press, 1990).

6. Physica D, Volume 38, which was also reprinted as Fractals in Physics, Essays in Honor of B.B. Mandelbrot on his 65th birthday, eds. A. Aharony and J. Feder (North Holland, 1989).

7. Fractals Volume 3 (September 1995), reprinted as Fractal Geometry and Analysis: The Mandelbrot Festschrift, Curaçao, 1995 eds. C.J.G. Evertsz, H.-O. Peitgen & R.F. Voss.

8. On the physics, a standard textbook is Fractals by J. Feder (Plenum, 1988).

4 Fractal Transformations

Michael Barnsley and Louisa Barnsley

A strange game of soccer is used to introduce transformations and fractals. Low information content geometrical transformations of pictures are considered. Fractal transformations and a new way to render pictures of fractals are introduced. These ideas have applications in digital content creation.

Born in 1946 to a literary household, Michael Fielding Barnsley escaped to pursue mathematics at Oxford (BA 1968) and theoretical chemistry at Wisconsin (PhD 1972). In 1985 he was appointed Professor of Mathematics at the Georgia Institute of Technology; and in 1987 he co-founded Iterated Systems Inc., which later licensed fractal image compression technology to Microsoft. His main research area is fractal geometry: he has lent to his name a number of patents, research papers, and books, supervised doctoral students and made one film. He is currently a professor at The Australian National University in Canberra.

N. Lesmoir-Gordon (ed.), *The Colours of Infinity: The Beauty and Power of Fractals*, DOI 10.1007/978-1-84996-486-9_4, © Springer-Verlag London Limited 2010

A fascinating soccer game – See Debbie kick!

Imagine that you are a great soccer player, with perfect ball control and perfect consistency. Wherever the ball is on the soccer pitch you can kick it so that it lands halfway between where it was and a corner. And the ball comes to a dead stop right where it lands. You can always do this.

That is how Alf, Bert, Charlie and Debbie are. They play soccer on the soccer pitch ABCD in Fig. 4.1. Debbie always kicks the ball when she gets to it first. She kicks it from X to the midpoint between D and X. See Debbie kick!

Alf acts in the same way, except that he kicks the ball halfway to A. And Bert kicks the ball halfway to B. You can guess where Charlie kicks the ball, when she gets to it first.

Who kicks the ball next is entirely random. It makes no difference where the players are on the pitch or who kicked it last. You can never reliably predict who is going to kick it next. The sequence of kickers might be determined by a random sequence of their initials: DABACBADAABCDCBACAADDBAC…

The game goes on forever.

To watch this awful game of soccer is rather like watching four chickens in a farmyard chasing after a bread crust. There is no team play, and no goals are ever scored. But at least no one eats the ball.

What actually happens to the ball is fascinating. Almost certainly it jumps around all over the pitch forever, going incredibly close to all of the points on the pitch. If you mark any little circle on the pitch, eventually the ball will hit the ground inside the circle. Sometime later it will do so again. And again and again. The soccer ball marks out the pitch, going arbitrarily close to every point on it. We say that the ball travels 'ergodically' about the pitch.

Alf, Bert, Charlie and Debbie represent 'transformations' of the soccer pitch. Alf represents the trans-

formation that takes the whole pitch into the bottom left quarter. Let ■ denote the soccer pitch. Then

$$\text{Alf}(\blacksquare) = \textit{Quarter A of Soccer Pitch,}$$

the quarter at the bottom left. Think of Alf 'kicking' the whole pitch into a quarter of the pitch.

Similarly Bert (■) = *Quarter B of Soccer Pitch*, the quarter at the bottom right. Also Charlie (■) = *Quarter C of Soccer Pitch*, and Debbie (■) = *Quarter D of Soccer Pitch*, the quarter at the top left.

These transformations actually provide an 'equation' for the soccer pitch:

$$\blacksquare = \textit{Alf}(\blacksquare) \cup \textit{Bert}(\blacksquare) \cup \textit{Charlie}(\blacksquare) \cup \textit{Debbie}(\blacksquare).$$

It says that that the pitch ■ is made of 'four transformed copies of itself'. It says that the pitch is the union of the four quarter-pitches, just as the UK is

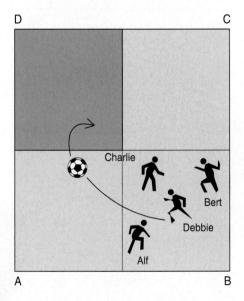

Fig. 4.1 above: Debbie has just kicked the ball halfway towards *D*. If the ball was at *X*, then it lands at the midpoint of the line segment *XD*.

the union of England, Northern Ireland, Scotland and Wales.

For us each player is a transformation or a function, providing a unique correspondence between each location on the pitch (from where the ball is kicked) and another location on the pitch (the point where the ball lands).

Charlie hurts her leg

What has all this got to do with fractals? Lots, as we shall see.

Suppose Charlie gets kicked in the shin and cannot play. Only Alf, Bert and Debbie kick the ball. Their sequence of kicks is still random, for example starting out in the order DBAABADBADDA BBAD...

The game begins with the 'kick-off', with the ball in the middle of the pitch.

Where now does the ball go? To find out we cover it with greeny-black ink. Now the ball makes a dot on the white pitch every time it lands. It will make a picture while the game is played.

Amazingly, the picture it makes, almost always, looks like the one in Fig. 4.2. This is called 'The Sierpinski Triangle ABD'. We denote it by ▲. With Charlie out of the game, the ball travels ergodically about ▲. Mark a small circle centred at any point on ▲. The ball will visit this circle over and over again.

The Sierpinski Triangle ▲ is a bona fide fractal. Notice how it is 'made of three transformed copies of

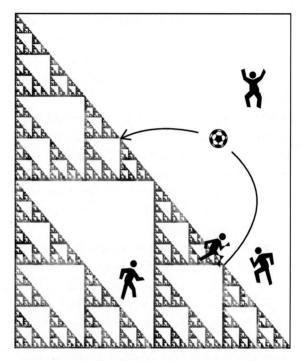

Fig. 4.2 above: Charlie hurts her leg and can't play for a while. The ball travels 'ergodically' on the Sierpinski Triangle ABD.

itself'. One copy lies in the top left quadrant, one in the lower left quadrant, and one in the lower right quadrant. It appears now that the soccer players 'kick' ▲ into smaller parts of ▲.

Our equation this time reads

$$\blacktriangle = Alf(\blacktriangle) \cup Bert(\blacktriangle) \cup Debbie(\blacktriangle).$$

This is the equation for m, the Sierpinski Triangle ABD.

What is amazing is that the transformations represented by the players, and this type of equation, define one and exactly one picture, the fern in this case, the Sierpinski Triangle in the previous case, and the whole soccer pitch in the first case.

The players change how they kick.

Charlie is back in the game.

Alf, Bert, Charlie and Debbie are fed up that they have not scored any goals. So they change the way they kick the ball.

Each player kicks in his or her own special way, methodically and reliably. Alf now kicks the ball so that it always lands in a certain quadrangle. Bert kicks the ball so that it lands in another quadrangle. Charlie and Debbie kick the ball into their own quadrangles. You can see the soccer pitch and the four quadrangles in Fig. 4.3.

Alf kicks straight lines into straight lines in this way. Let P, Q and R be three points that lie on a straight line. He kicks the ball from P to $Alf(P)$, from Q to $Alf(Q)$ and from R to $Alf(R)$. Then $Alf(P)$, $Alf(Q)$, and $Alf(R)$ lie on a straight line! For example, if the ball is on one of the sidelines of the soccer pitch, Alf kicks it to land on one of the sides of his quadrangle. If the ball lies at the centre of the soccer pitch, Alf kicks it so that it lands at the intersection of the two diagonals of his quadrangle. Alf is a precision kicker.

If Alf kicks the ball from two different points, it lands at two points closer together than the starting points. We say that Alf represents a 'contractive' transformation.

The other players act similarly; the only differences are the quadrangles that they kick to.

Where does the ball go this time? The selection of the order in which the players kick the ball is again random. The game goes on for eons. The players never get bored or tired. They are immortal. And the ball marks green or black points on the white pitch wherever it lands, after the first year of play. The resulting pattern of dots forms the fern in Fig. 4.3. We call this fern F. The soccer ball travels ergodically on F.

Our equation this time reads

$$F = Alf(F) \cup Bert(F) \cup Charlie(F) \cup Debbie(F).$$

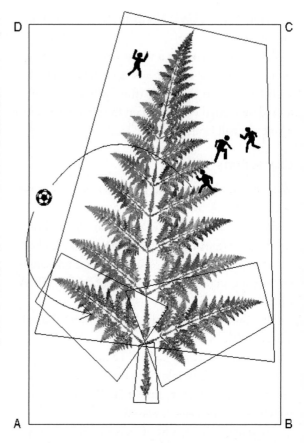

Fig. 4.3 above: Each soccer player now represents a projective transformation. One transformation corresponds to each of the quadrangles inside the pitch *ABCD*. The places where the ball lands make a picture of a fern. The ball travels 'ergodically' upon the fern.

The fern F is the union of 'four transformed copies of itself'.

What is amazing is that the transformations represented by the players, and this type of equation, define one and exactly one picture, the fern in this case, the Sierpinski Triangle in the previous case, and the whole soccer pitch in the first case. Change the way the players kick the ball and you will change the picture upon which eventually the ball 'ergodically' travels.

Infernal football schemes (IFS)

Many different fractal pictures and other geometrical objects can be described using an IFS. The letters really stand for Iterated Function System but here we pretend they mean Infernal Football Scheme.

An IFS is made up of a soccer pitch and some players each with their own special way of kicking. Each player must always kick the ball from the pitch to the pitch according to some consistent rule. And each player must represent a contractive transformation, must 'kick' the pitch into a smaller pitch. We continue to call the players Alf, Bert, Charlie and Debbie, but there may be more or less players.

Then there will always be a unique special picture, a 'fractal', a collection of dots on the white pitch, which obeys the equation

$$Fractal = Alf(fractal) \cup Bert(fractal) \cup Charlie(fractal) \cup Debbie(fractal)$$

We call this picture a fractal, but it might be something as simple as a straight line, a parabola, or a rectangle. This picture can be revealed by playing random soccer as in the above examples.

An example of a fractal made using an IFS of three transformations is shown in Fig. 4.4.

In this way many fractals and other geometrical pictures can be encoded using a few transformations. Once one knows an IFS for a particular fractal one knows its secret. One knows that despite its apparent visual complexity, it is really very simple. One can make it and variations of it, over and over again. One can describe it with infinite precision.

Given a picture of a natural object, such as a leaf, a feather, or a mollusc shell, it is interesting to see if one can find an IFS that describes it well. If so, then one would have an efficient way to model and compare some biological specimens.

Fig. 4.4 left: Fractal made with three projective transformations. It is rendered in black and shades of purple. Can you spot the transformations?.

Fig. 4.5 left: Two circlepre-serving projective transfor-mations of a picture of an Australian heath.

Fig. 4.6 below: Two different ellipsepreserving projective transformations of a beech leaf.The straight lines along which the veins nearly lie are preserved.

Geometrical transformations

Simple transformations

So far we have shown that to understand fractals one needs to understand transformations. But which transformations?

Transformations can be very complicated. They may involve bending one part of space, squeezing another, and be expressed using elaborate formulas that take pages to write down.

But one of the goals of fractal geometry is to describe pictures of natural objects in an efficient manner. Clearly, if one makes a description of a fern using an IFS, and the transformations that are used are very complicated, then little is gained in the way of simplification. So we seek simple transformations – ones that are easy to write down, explain, and understand.

One source of simple transformations is classical geometry, which involves the study of invariance properties of collections of transformations. For example, Euclidean geometry studies properties of geometrical pictures that remain unchanged when elementary displacements and rotations are applied to them. The distance between a pair of points is invariant under a Euclidean transformation. So is the angle between a pair of straight lines.

Similarity geometry involves the transformations of Euclidean geometry as well as similarity transformations, so called because they magnify or shrink pictures by fixed factors. Many well-known fractals may be expressed with similarity transformations, for example the Sierpinski Triangle ▲ and the soccer pitch ■. But projective geometry provides a much richer simple set of transformations for describing natural shapes and forms.

Projective transformations

Projective transformations are of the type represented by Alf, Bert, Charlie and Debbie when they started kicking the soccer pitch into quadrangles. Given any pair of quadrangles, one can always find a projective transformation that converts one into the other, even making the corners go to specified corners.

Projective transformations arise naturally in optics, in explaining perspective effects, and play an important role in modern physics. They seem to appear naturally when one searches for order and pattern in the arrangements of matter and light in the natural physical world.

They are indeed natural in the following way. Suppose you take a wonderfully sharp photo of a tree full of flat leaves, some bigger, some smaller, but all of the same shape. Then all of the whole leaves in the photo will be (almost) projective transformations of one another.

When you watch television from a difficult angle, the images that fall on your retinas are in effect projective transformations of what they would have been if you viewed face-on. But, within reasonable limits, the mind/eye system copes with the distortion. 'Recognisability' is an invariance property of projective transformations.

Projective transformations have the property that they often transform images of plants and leaves into recognizable images of plants and leaves. This is illustrated in Figs. 4.5 and 4.6. Note how the straight lines of the veins in the beech leaf are transformed into other straight lines in Fig. 4.6.

Images of the real world contain much repetition. Often nearby leaves look similar for biological and physical reasons. And the local weather pattern seems to clump clouds into regions of similar looking ones. This similarity and repetition may be specified with projective transformations.

Michael Barnsley in the 'dream'sequence from 'The Colours of Infinity' film:'I woke up in the morning and knew that I'd discovered it.This was the total secret to fractal image compression:how to automatically look at a digital picture … and how to turn it into (a) a formula and (b) an entity of infinite resolution. So the goal is now to be able to capture this Fire of Prometheus, if you like, this fractal wonder, put it in a box and being able to make this available to everyone.'

Projective transformations take points to points and straight lines to straight lines. Even more remarkable, they map conic sections into conic sections. That is, if you make a picture of circles, ellipses, parabolas, hyperbolas and straight lines, then apply a projective transformation, the resulting picture will also be made of these same shapes. That is not to say that circles are transformed to circles, ellipses to ellipses, parabolas to parabolas, or hyperbolas to hyperbolas.

Are the coloured circular and elliptical cells on the wings of some butterflies more easily recognized by other butterflies, or predator species, because of this invariance?

Mobius transformations

Mobius transformations are another type of transformation that is 'simple'. They are often used to describe fractals and, in a different way than projective transformations, seem to have some natural affinity with real world images. They have the remarkable property that they transform any circle into either a circle or a straight line. This is illustrated in Fig. 4.7. They also preserve the angles at which lines in pictures cross, as can be seen by examining the bike frames in Fig. 4.8.

In certain situations they transform patterns of fluid motion, represented by streamlines, into other possible fluid motion patterns. They also transform pictures of fish into other pictures of fish, as illustrated in Fig. 4.8.

Mobius transformations are the basic elements of hyperbolic geometry. They were used by Escher in some of his graphic designs, including ones with natural elements such as fish.

In Fig. 4.9 we illustrate the Circumscribed Fish Theorem. This is one of many such observations. It illustrates that geometry applies not only to triangles, circles, and straight lines, but to all sorts of other pictures as well.

Fig. 4.7 left: A single Mobius transformation is applied over and over again to a picture of a person on a bike. The images are massively distorted one from another, but the wheels are all round, except near the edges of the picture, where some precision has been lost. Also angles are preserved. Each bicycle frame is a curvilinear triangle with the same three angles.

Fig. 4.8 above: The same Mobius transformation is applied over and over again to a single fish, to produce this double spiral of fish. Notice that although the fish are massively distorted, they all look fish-like.

Fig. 4.9 below: Illustration of the Circumscribed Fish Theorem. Although the fish in Fig. 4.8 look quite various, they have the following property: Draw the smallest circle around each fish, such that the circle touches the fish in at least three points; then each fish touches its circle with the same parts of its body.

The cost of describing transformations

Even 'simple' transformations can be complicated if they involve 'constants' that require lots of digits to express them accurately. To explain this point, let us look briefly at some 'formulas' for simple transformations. The details of these formulas, other than the fact that they contain 'constants', need not concern us.

Transformations in two-dimensional space may be represented using Cartesian coordinates (x, y) to represent points. A projective transformation can be expressed with a formula such as

$$\left(\text{Alf}(x,y) = \frac{ax+by+c}{gx+hy+1}, \quad \frac{dx+ey+f}{gx+hy+1} \right)$$

where a, b, c, d, e, f, g, h, and k, are numbers, the 'constants', such as a = 1.023, b = 7.1, c = −0.00035, d =1 00, f = 9.1, g = 34.9, and h = 17.3. Similarly, a Mobius transformation can take the form

$$Bert(x,y) = \left(\frac{a+(\sqrt{}-1)b+(c+(\sqrt{}-1)d(x+(\sqrt{}-1)y)}{e+(\sqrt{}-1)f+(g+(\sqrt{}-1)h)(x+(\sqrt{}-1)y)} \right)$$

which uses complex arithmetic and also uses eight constants.

If we know that each constant is an integer between −128 and +128, which can be expressed using one byte of data (since 2^8 = 256) then each of these transformations requires 8 bytes of information to express it, one byte for each constant. These are in, an obvious way, 'simpler' transformations than ones in which each constant requires two bytes of information. And both of these possibilities are much simpler, that is, able to be expressed much more succinctly, than if each constant were a decimal number with random digits, such as a = 1.79201434953…, going on forever.

Now one might say that all of these extra digits are without significance. But in fractal geometry they are very significant, because fractal geometry is about details! Tiny changes in the constants will usually lead to tiny changes in a fractal built using the transformations.

Transformations can be very complicated. They may involve bending one part of space, squeezing another, and be expressed using elaborate formulas that take pages to write down. But one of the goals of fractal geometry is to describe pictures of natural objects in an efficient manner.

But when the fractal is under the microscope, so to speak, and one is zoomed in to look at fine detail, and a tiny change is made in a coefficient, the part of the fractal one is looking at may completely disappear – not only has its form changed, but it has moved out of the field of view.

For the application of fractals to image compression, for example, it is important that the transformations can be expressed succinctly, and that the constants involved do not require lots of digits. We say that such transformations have 'low information content'.

One of the important features of fractals and other geometrical pictures is that they are simple to describe. Thus it is appealing to use low information content transformations, quite generally.

More soccer: Fractal transformations are discovered.

Alan, Brenda, Celia and Doug start a second game

We can use fractal soccer, with simple projective 'kicks', to make a new kind of transformation. We call these new transformations 'fractal transformations'. They too are of low information content. But they can transform pictures in very surprising ways, very differently from projective and Mobius transformations.

In Fig. 4.10, two games of soccer are played at the same time. The game on the left is the same as in Fig. 4.1 above. But in the game on the right, Doug kicks the pitch into the small rectangle at the top left, while Brenda kicks into the large rectangle at the bottom right. Similarly, Celia kicks towards C and Alan kicks towards A, but the quadrangles that

they kick to are of different dimensions than in the first game.

Alan, Brenda, Celia and Doug are copycats. They watch the game on the left. When Alf kicks the ball, Alan kicks the ball in his game; when Bert kicks the ball, Brenda kicks the ball; when Charlie kicks the ball, then so does Celia; and when Debbie kicks the ball, so does Doug – he's been watching her closely. But of course Alf, Bert, Charlie and Debbie stay on the pitch on the left, while Alan, Brenda, Celia and Doug stay to their soccer pitch on the right.

Now put a picture on the soccer pitch on the left, a great big one. This is the 'Before' picture. To illustrate this, there is a big red and green fish painted on the left-hand pitch in Fig. 4.11.

Let the game begin. Then after each pair of kicks, one on each pitch, a dot is painted on the right-hand pitch at the spot where the ball has landed, in the same colour as the point on the left-hand pitch where the ball on that pitch has landed. The result, after thousands and thousands of kicks, is shown in Fig. 4.11, on the right-hand pitch. This is the 'After' picture.

The After picture is an amazingly deformed version of the Before picture, stretched greatly in some places and only a little in others. We call this a fractal transformation.

But the transformation between the Before and After pictures is fundamentally no more complicated than the transformations that are used to make it, the transformations represented by the players.

Fig 4.10 right: Alan, Brenda, Celia and Doug start up a second game. They are copy-cats: Doug kicks the ball whenever Debbie does, Alan kicks the ball whenever Alf does, Celia kicks when Charlie does, and Bert copies Brenda. But they kick the ball a bit differently!.

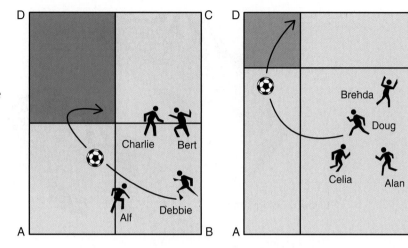

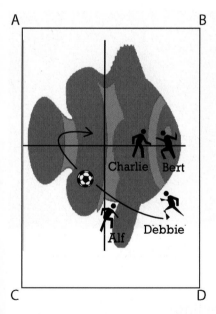

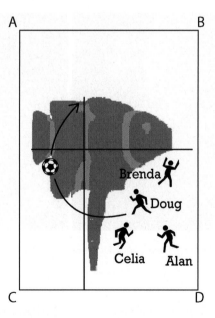

Fig. 4.11 left: The fish is transformed by the two soccer games.

Although the player transformations are quite smooth and regular, the fractal transformation is non-uniform and irregular.

In Fig. 4.12, we show a prettier fish, before applying a fractal transformation to it. In Fig. 4.13, we show the same fish after transformation. Another before and after pair is shown in Fig. 4.14. Such effects clearly have applications in digital content creation.

In Fig. 4.15, we show a before-and-after pair of pictures of Australian heath flowers. It is interesting to compare this figure with Fig. 4.5, where the two

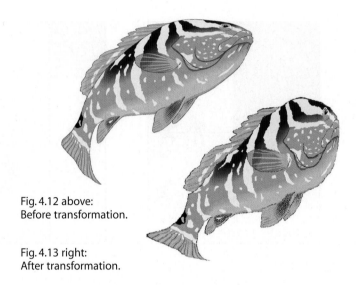

Fig. 4.12 above:
Before transformation.

Fig. 4.13 right:
After transformation.

Fig. 4.14 above: These two pictures of leaves and sky are related via a fractal transformation.

images are related by a circle-preserving projective transformation. In the present case the images are related by a rectangle-preserving fractal transformation (the rectangular picture frame is preserved). Under projective transformation, points that are collinear are mapped into collinear points. Under the present fractal transformations collinear points *parallel to the picture frames are preserved.*

Colour stealing

Essentially the same algorithm to the one we have described in the previous section may be applied to render rich colouring to diverse IFS fractals. Here we show how a fern is coloured by this new algorithm. See Fig. 4.16. The main difference is that on the right-hand pitch the IFS that makes the fractal fern is used.

Fig. 4.15 above: These images of Australian heath are related by a rectangle-preserving fractal transformation. Compare with Fig. 4.5.

Fig. 4.16 left: On the left-hand soccer pitch with the colourful photo on it, Alf, Bert, Charlie and Debbie play a game of random soccer. The players in the game on the right are Alan, Brenda, Celia and Doug. They kick the ball into quadrangles, as in Fig. 4.3. Alan kicks the ball when Alf does, Brenda kicks it when Bert does, and so on. Each time after both balls have been kicked, the spot where the ball lands in the right-hand game is marked with a dot the same colour as the point where the ball lands in the left-hand game. The result is a painted fractal fern.

On the left-hand soccer pitch with the colourful photo on it, Alf, Bert, Charlie and Debbie play a game of random soccer as in Fig. 4.1. Each player simply kicks the ball to the quarter pitch labelled with his or her initial. The players in the game on the right are Alan, Brenda, Celia and Doug. They kick the ball into quadrangles, as in Fig. 4.3. Alan kicks the ball when Alf does, Brenda kicks it when Bert does, Celia kicks when Charlie does, and Doug kicks when Debbie does.

A while after kick-off, each time after both balls have been kicked, the spot where the ball lands in the right-hand game is marked with a dot the same colour as the point where the ball lands in the left-hand game. The result is a fractal fern, painted with the colours of the picture on the left.

Fig. 4.17 contrasts two copies of the same fern coloured by a fractal transformation of two different pictures, samples of which are shown at left. Notice that there need be no particular relationship between the size of the picture from which the colour is stolen and the target image, the fern in this case, that is painted with the stolen colours.

Comments, background references and further reading

The ideas of fractal transformations and colour stealing using random iteration, the main topics of this chapter, are, so far as we know, entirely new and are presented for the first time here. What is actually going on in both cases is that a mapping is set up between two IFS attractors using the underlying code space, which is the same for both IFSs.

This means that a fractal transformation between two 'just-touching' IFS attractors is very nearly con-

Fig. 4.17 right: The same fractal fern is rendered using two different input images, shown on the left.

tinuous, which explains why the colourings of the fern, for example, have a nice consistency from one frond to the next and do not vary too abruptly.

The random soccer game is a novel way of presenting geometrical transformations, the random iteration algorithm and IFS theory. Our goal has been to minimize the use of formulas and to try and rely on geometrical intuition and non-mathematical wording. The random iteration algorithm was first described formally, in the context of fractal imaging, in [1], although the seeds of this idea are mentioned in the early work of Mandelbrot, [9], p. 198. This algorithm is also known as the 'Chaos Game', but we think it may attract a wider audience if it is explained in terms of soccer.

The mathematical theory of IFS was originally formulated by John Hutchinson [6]. It was popularized and developed by one of us and co-workers as well as many others, see for example [5] and [8]. You can read about the application of IFS to image modelling, how to make fractal ferns and leaves, and about the underlying code space, in [2]. The application of IFS to image compression is described in [3] and in [7]. A lovely book about fractals made with Mobius transformations is [10].

The future holds another exciting discovery, which you may read about in [4] and also hopefully in 2004 in a book entitled Superfractals.

References

1. M.F. Barnsley and S. Demko, 'Iterated Function Systems and the Global Construction of Fractals', *R. Soc. Lond. Proc. Ser. A Math. Phys. Eng. Sci.* 399 (1985), 243–275.

2. M.F. Barnsley, *Fractals Everywhere*, Academic Press, New York, 1988.

3. M.F. Barnsley and L.P. Hurd, *Fractal Image Compression*, A.K. Peters, Boston, MA, 1993.

4. M.F. Barnsley, J.E. Hutchinson and Ö. Stenflo, 'A New Random Iteration Algorithm and a Hierarchy of Fractals', Preprint, Australian National University, 2003.

5. K. Falconer, Fractal Geometry – *Mathematical Foundations and Applications*, John Wiley & Sons, Chichester, 1990.

6. J.E. Hutchinson, 'Fractals and Self-Similarity', *Indiana. Univ. Math. J.* 30 (1981), 713–749.

7. N. Lu, *Fractal Imaging*, Academic Press, San Diego, 1997.

8. H.O. Peigen and D. Saupe, *The Science of Fractal Images*, Springer-Verlag, New York, 1988.

9. B.B. Mandelbrot, *The Fractal Geometry of Nature*, W.H. Freeman and Company, San Francisco, 1983.

10. D. Mumford, C. Series and David Wright, *Indra's Pearls*, Cambridge University Press, Cambridge, 2002.

5 Fractal Limits

The Mandelbrot Set and the self-similar tilings of M.C. Escher

Will Rood

Will Rood, a pioneer fractal animator, says many people have asked him about the apparent similarities between the M-set and the fantasy lizards and wheelie animals of iconic Dutch artist M. C. Escher. This chapter develops the theme of art at the fractal limits and styles of colouring the M-set by exploring the meeting of Mandelbrot and Escher.

Mathematician Will Rood discovered fractals in 1984, only to find that Benoît Mandelbrot already knew about them ten years earlier. Undeterred, Rood went on to graduate in pure mathematics at Cambridge, and was awarded for his work on transfinite set theory. He now writes software and makes films and videos. He has collaborated with Nigel Lesmoir Gordon on many projects, including the original film *The Colours of Infinity*.

Fig. 5.1 right: The meeting of Mandelbrot and Escher.

N. Lesmoir-Gordon (ed.), *The Colours of Infinity: The Beauty and Power of Fractals*, DOI 10.1007/978-1-84996-486-9_5, © Springer-Verlag London Limited 2010

What is the Mandelbrot Set?

It's been called the most complex shape known to man. A lifetime would not be enough to get to the bottom of it. And yet it is generated by a formula of surprising simplicity. In principle it could have been discovered at any time in human history. However, in practice we had to wait until the invention of the silicon chip, for visualizing the Mandelbrot Set involves applying this simple formula, over and over and over again.

Benoît Mandelbrot discovered his eponymous Set while delving into the almost forgotten work of his uncle's teacher, Gaston Julia. In the 1920s Julia discovered a whole class of strange and beautiful shapes, now known as Julia Sets. However, Julia never actually saw a Julia Set. Without computers, he could only have a vague idea of their true form.

Working at IBM in the late 1970s, Mandelbrot was one of the first scientists with enough computing power at his disposal to generate Julia Sets. He created many stunning images of these shapes, but what he really wanted to know was the overall pattern behind the whole family of Julia Sets. He decided to make a map of them, which map is now called the Mandelbrot Set.

The Mandelbrot Set is a strange shape, and the closer you look, the stranger it becomes. It's been called the thumbprint of God and the Creator's calling card – clear evidence of a deep underlying harmony and unity in nature.

A deeper journey

The main body of the Mandelbrot Set consists of a cardioid, or heart-shaped core, surrounded by infinitely many circular buds. Each bud is surrounded by a further infinity of smaller buds, and, at the end of each of these chains of buds, a spiral frond, sometimes lacy and floral, sometimes straight and spiky. The fronds, which comprise the boundary of the Mandelbrot Set, actually consist of infinitely many miniature copies of the whole shape, joined together by bifurcating threads of ever-smaller miniatures.

In response to this continual branching, these fronds are also called dendrites, from the Greek for tree. The name conjures up associations with the straggly branching receptors of nerve cells in our brains, which are also called dendrites. This is no accident: evidently the functionality and processing power of neurons derive from their richly entwined fractal structure.

The formula

The Mandelbrot Set looks very complicated, and yet it is generated by a very simple rule:

$$Z \rightleftharpoons Z^2 + c$$

The arrow can be read as 'goes to' or 'becomes', for what this rule represents is a transformation of two-dimensional space; the letters z and c by convention indicate generic points in this space, with z being variable and c constant. In other words, the rule transforms the point z to another point in the space, while leaving c unchanged. This two-dimensional space inhabited by z and c, the home of the Mandelbrot Set and Julia Sets, is central to mathematics, from quantum mechanics to number theory.

The complex plane

In the early sixteenth century mathematicians such as Gerolamo Cardano discovered that certain numerical problems are easier to solve if we pretend that every number has an imaginary part, with special properties. Adding these imaginary parts makes no difference to the real part of the number, but multiplying two imaginary parts produces a negative number. Gradually it became apparent that these complex numbers are not only a natural extension of our normal numbers, they also encapsulate the geometry of two dimensions in a beautifully compact form.

Mapping the real and imaginary parts of numbers onto the axes of a graph, the arithmetic operations become elementary geometrical transformations. (see Fig. 5.2) Addition becomes translation; multiplication becomes rotation. The function $z^2 + c$ can be interpreted as stretching and wrapping the plane twice around itself.

The Mandelbrot Set emerges when we apply this rule over and over again, taking the outcome of one transformation as the input for the next. Like the clues in a treasure hunt, for each point the rule gives us the location of the next point in the sequence.

The map is the treasure

The process of repeatedly applying a rule is called iteration. When we do this, or, more likely, program a computer to do this, one of two things can happen: either the point z gets very big, that is, very far from the origin 0, or it doesn't.

That much is obvious. Furthermore, once z reaches a certain size, it keeps on getting bigger and bigger, in which case we say the point z goes to infinity, otherwise z remains bounded. What is less obvious is that colouring these points, according to whether or not they go to infinity, will produce pictures of enormous complexity and aesthetic appeal.

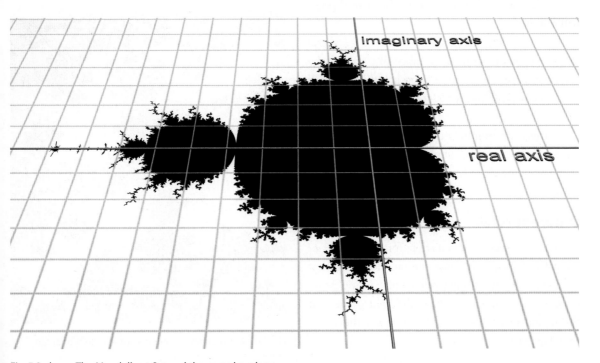

Fig. 5.2 above: The Mandelbrot Set and the complex plane.

In these maps, black represents points that remain bounded.

If we iterate $z = z^2 + c$ for an array of different z values, keeping c fixed, we build up a picture of the Julia Set for that value of c. (Fig. 5.3)

If we iterate the rule for one particular value of z, and an array of different c values, we get a picture of the Mandelbrot Set. (Fig. 5.4)

The critical point

The particular value of z that we start with is a special one: $z = 0$. It is what is called the critical point for the mapping $z^2 + c$.

What this means is that the behaviour of the point $z = 0$ indicates the general effect of this mapping on the entire plane. If this point goes to infinity, then so do almost all points, and the Julia Set is a disconnected dust; otherwise the Julia Set is connected, and if there is an attractive cycle, a set of points that pulls others towards it, the critical point will find it.

If we use a different starting point, say $z = -1$, and iterate this point for a whole range of c values, the resultant picture is a bizarre hybrid, with no overall cohesion, looking like a M-set with chunks missing. (Fig. 5.5)

So there are infinitely many different Julia Sets, one for each value of c, but only one Mandelbrot Set, as the mapping $z^2 + c$ has only one critical point.

Different rules

So what is special about this formula, $z^2 + c$? Actually, nothing. It's just the simplest rule that generates any interesting behaviour. Using more complicated formulae produces different shapes but the boundary details remain very much the same. (See Fig. 5.6).

Fig. 5.3 above: Map of z-plane showing Julia Set in black.

Fig. 5.4   below: Map of c-plane with Mandelbrot Set in black.

Escape time

There are many ways of colouring the exterior of the M-set that help us visualize how this mapping $z^2 + c$ transforms the complex plane. The simplest is to colour each point according to how many iterations it takes to reach a cut-off region $|z| > R$, where $R \geq 2$. This is called the escape time method, since it measures the time for a point to 'escape' into the cut-off region: any points that fail to escape after a given time are assumed to belong to the Mandelbrot Set, and are traditionally coloured black.

The colours we assign to the outside of the M-set are quite arbitrary: they serve to illustrate the level sets of points that escape in equal time. (Fig. 5.7)

The level sets are loops, which each make one complete circuit of the M-set. Alternating black and white gives striped bands, which, near the boundary of the M-set, form highly convoluted squiggles reminiscent of the op-art of Bridget Riley. (Fig. 5.8)

Smooth shading gives a clearer view of the edge of the M-set (Fig. 5.9), and more advanced methods such as distance estimation reveal a clearer picture still of this intricate boundary, (Fig. 5.10) but the striped black and white bands accentuate the behaviour of points farther away from the M-set.

Continuous escape time

Like the contours on a map, the exact position of these bands is arbitrary, determined in this case by the size of the cut-off region. A more natural representation of the exterior of the M-set can be achieved using the continuous escape time method. This uses the escape time as a starting point, and then adds a correction factor between 0 and 1, depending on how far into the escape region the point got.

If z is the final point in our sequence, and R is big enough, we know that the following are roughly true

Fig. 5.5 above: Map of c-plane for initial point $z = -1$.

Fig. 5.6 below: Detail of Julia Set for 4th order quotient map

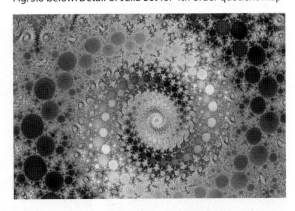

(where $|z|$ is the modulus of z, the distance from z to 0):

$$R < |z| \leq R^2$$
$$\log R < \log|z| \leq 2\log R$$
$$\log_2\log R < \log_2\log|z| \leq 1 + \log_2\log R$$
$$0 < \log_2\log|z| - \log_2\log R \leq 1$$

So we can define the continuous escape time cet as:

$$cet = \text{escape time} + \log_2 \log R - \log_2 \log |z|$$

Fig. 5.7 above: Level Sets of $z \rightleftharpoons z^2 + c$.

For large R this function is relatively smooth on the entire exterior of the M-set. (Fig 5.11) The continuous escape time gives precise information about the rate of escape of each point, but since it uses the modulus of the final point and not its direction, provides no picture of the shape of the trajectory of the point. The smooth contours suggest a steady outward flow, but this is illusory. The contours may flow smoothly into each other but the points on them jump all over the place, since the mapping $z^2 + c$ wraps the entire plane twice around itself. (Of course, this dynamic takes place in the z-plane, not the c-plane shown in most of these pictures, but remarkably these methods work just as well whether applied to the Mandelbrot or Julia Sets.)

Escape angle

Plotting the escape angle, the angle the final point makes with the real axis, for each point that escapes, demonstrates this. (Fig. 5.12)

Fig. 5.8 above: Level sets.

Fig. 5.9 above: Smooth shading.

Fig. 5.10 above: Distance estimation.

Fig. 5.11 above: Three-dimensional plot of continuous escape time.

The gradient flows perpendicular to the escape time contours, which are still visible as discontinuities in the escape angle. Each band has been wrapped around itself once more than its predecessor, giving it twice as many bright stripes. (Fig. 5.13)

Taken together, the continuous escape time and the escape angle provide an orthogonal grid – they intersect each other only at right angles – since the real axis cuts the escape region at right angles and the mapping $z^2 + c$ is conformal: it preserves angles.

This implies that pictures mapped onto this grid will be relatively undistorted. These pictures reveal where each escaping point ends up. Points mapped to the same part of the picture end up in the same place. (Fig. 5.14 a and b)

Self-similar tessellations

M.C. Escher, the undisputed master of tessellated (covering a surface with closely fitting pieces) art, often drew inspiration from mathematical sources. The geometer Donald Coxeter introduced him to hyperbolic tessellations, tilings of the strange non-Euclidean space discovered by Bolyai and Lobachevsky around 1820 where through each point there are many parallels to a given line. When

Fig. 5.12 above: Map of escape angle.

Fig. 5.13 right: Detail of escape angle map.

viewed in Euclidean space, these seem to contain cascades of ever-smaller tiles, although, in the hyperbolic space represented, all these tiles are exactly the same size. (Fig. 5.15)

The apparent shrinking can be thought of as a perspective effect of the mapping, analogous to the apparent diminishing of size on a receding plane. (Fig. 5.16)

These hyperbolic tilings solved a persistent problem for Escher: how to represent infinity in a closed form, and he used them in a series of woodcuts, Circle Limit I – IV.

As well as these hyperbolic tilings, Escher also developed methods for generating self-similar

Euclidean tilings, which repeat in one direction while shrinking in another. Coxeter was not overly impressed with these:

> ... not very interesting. The circle limits are much more interesting, being non-Euclidean.[1]

However, it seems as though Escher had actually come up with the archetypal self-similar Euclidean tiling, which, by a magnificent stroke of luck, is perfect for mapping onto the exterior of quadratic

---

[1] Bruno Ernst, The Magic Mirror of M.C. Escher (Tarquin Publications, 1985).

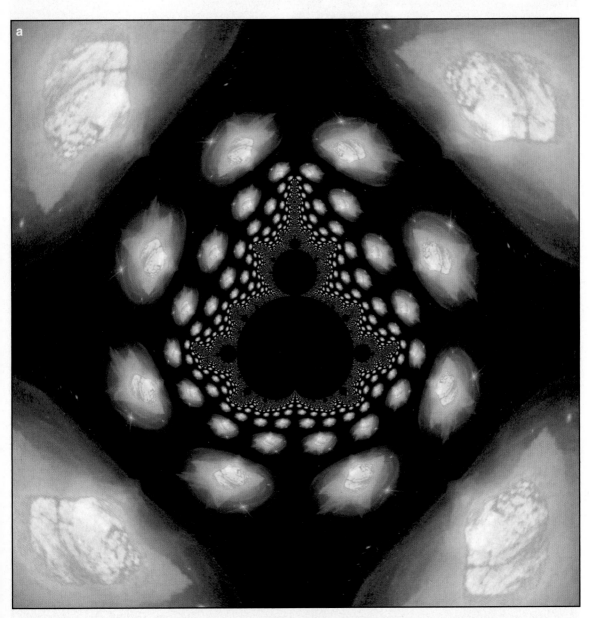

Fig. 5.14 a above and b overleaf: These pictures indicate the escape route of points outside the Mandelbrot Setb above.

Fig. 5.14 b above.

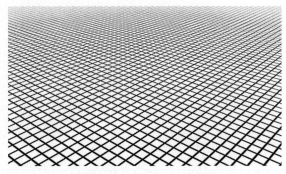

Fig. 5.16 above: Perspective effect: in the space depicted, all the squares are the same size, although in this projection some appear larger than others.

Fig. 5.15 above: Tiling of hyperbolic space[2].

fractals, allowing the creation of tessellations with fractal limits. (see Fig. 5.17)

Escher's reptiles can be seen to follow lines of flow, based upon the escape time and escape angle contours. The same patterns occur in the foliage that surrounds the miniature copies buried deep within the M-set. The smaller the miniature, the more detailed the foliage. (Fig. 5.18)

---

[2] Don Hatch, www.hadron.org

Fig. 5.17 above: Fractal limit: Escher's lizards surround the Mandelbrot Set

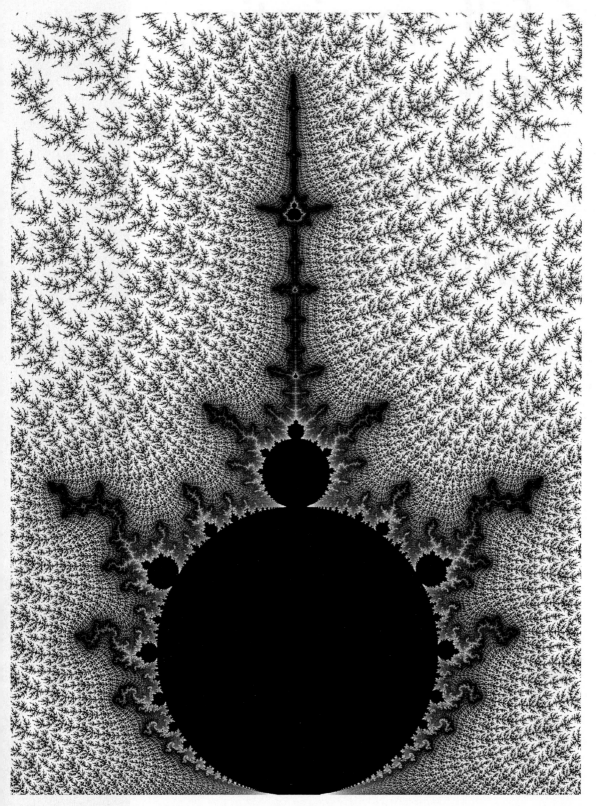

Fig. 5.18 right: The main spike of a miniature, deep within the Mandelbrot Set

6 Self-organization, Self-regulation, and Self-similarity on the Fractal Web

Gary William Flake and David M. Pennock
Yahoo! Research Labs

The authors begin by modelling the World Wide Web as an eco-system, which reflects an intimate coupling of people, programs, and pages. Viewing the Web from a variety of scales and view-points, from macroscopic to microcscopic, it is evident that users, authors, and search engines all influence one another to yield an amazing array of self-organization, self-regulation, and self-simi-larity. Ultimately, the Web's organization is intimately related to the complexity of human culture and to the human mind, and it is this subtle relationship between humanity and the Web that is responsible for the Web's amazing properties.

Gary Flake is a leading researcher in Web analysis and modelling, and author of *The Computational Beauty of Nature: Computer Explorations of Fractals, Chaos, Complex Systems, and Adaptation* (MIT Press, 1998). He is presently head of Yahoo! Research Labs in Pasadena, California.

David Pennock is a Seni... tist at Yahoo! Research ... with Gary Flake at NEC. ... Pennsylvania State Univ... a number of patents rel... commerce and the Wek... has received significant ... e-market companies an...

N. Lesmoir-Gordon (ed.), *The Colours of Infinity: The Beauty and Power of Fractals*,
DOI 10.1007/978-1-84996-486-9_6, © Springer-Verlag London Limited 2010

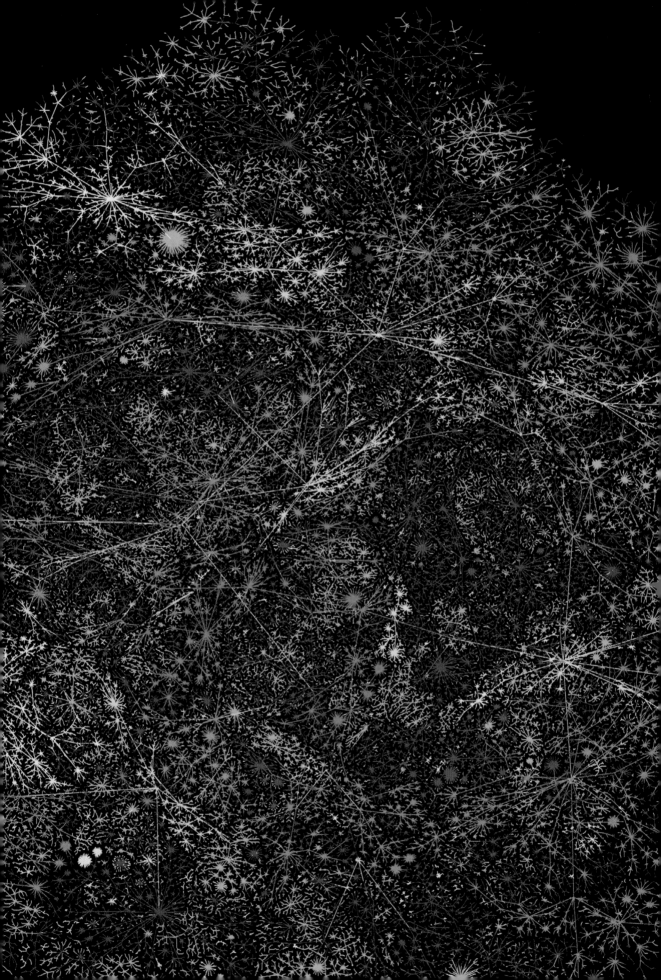

The Web as an Ecosystem

The World Wide Web is a digital entity like no other. Over the course of roughly fifteen years – and at an exponentially increasing rate – the Web has managed to capture, collect, organize, and connect a stunning amount of humankind's collective knowledge. It now reflects almost every aspect of our collective culture: from the peace prize to pornography; from academia to e-commerce; and from the mega-corporation to the personal home page. Though the Web is certainly a unique object in the history of the world, at its heart the Web is a social creation, and so perhaps it is not surprising that many of the Web's properties mimic those of nearly every other social and biological entity. The Web is in a very real sense an ecosystem, and as such can be viewed from a number of different perspectives spanning the microscopic to the macroscopic, with each vantage point showing an astonishing amount of complexity.

Natural ecosystems derive much of their complexity from a vast number of interdependencies: predators consume prey; individuals compete for the opportunity to reproduce; symbiotes cooperate with other species for improved viability; and the expired biomass from all organisms ultimately fuels the microbes at the lowest level of the food chain. In this way, an ecosystem is endlessly circular, with chains of dependencies streaming between individuals and species.

Fig. 6.1 previous page: A map of part of the Internet's topology, updated March 2004, illustrating the macroscopic structure of the Web and the apparent fractal nature of link connectivity. Points correspond to distinct Internet addresses of computers on the Internet; lines correspond to the connections between computers.

Data and visualization courtesy Bill Cheswick and Hal Burch of Lumeta Corporation. Lumeta is a pioneer in analyzing and securing corporate networks, http://www.lumeta.com. Reprinted by permission.

We say that an ecosystem's state is recursive because of the circularity of the ecosystem's dependencies. The future of every creature is intimately coupled to the present state of every other member of the ecosystem. As a result, the life cycle of a single individual as well as the evolution of an entire ecosystem are both tremendously complex precisely because each is a function of the other.

The circular dependencies of the Web are rich as well. Web authors attempt to build pages that a target audience of users will value, and the authors add value by supplying a mixture of content and hyperlinks (or more simply, links) to other valuable pages. Hence, one instance of recursion on the Web is that valuable pages tend to accumulate incoming links, and pages can become more valuable by linking to other valuable pages. The subtlety of the Web's recursion partially hinges on the circular influences that authors and users have on one another, each taking actions that are influenced by the other. To complete the analogy between the Web and natural ecosystems: the behaviours of individual authors or users as well as the evolution of the entire Web are tremendously complex precisely because each is a function of the other.

Throughout this chapter, we will use the analogy between natural ecosystems and the Web to better explore the Web's fractal properties and from whence they come. We will focus on three different vantage points: the microscopic level of the individual author or user (single organism), the intermediate level of the Web community (the niche or species), and the macroscopic level of the entire Web (the entire ecosystem or biosphere). But first, we will step back from the Web completely to examine its origin and evolution.

A Birds Eye View of People, Programs, and Pages

Before we dissect the Web in terms of scale, it is valuable to stand back and take a look at the Web's evolution in the broader context of human behaviour. Doing so will allow us to better understand and appreciate how the different scaling properties of the Web relate to one another and what external forces drive the dynamics of the Web. For this discussion, we will focus our attention on how users (people), search engines (programs), and Web sites (pages) impact one another.

At any moment in time, one can (in theory) measure the number of users that view a page over some period, the likelihood that a page will be the result of a typical query sent to a search engine, and the number of links that point to a particular page from other Web pages. Let's refer to these three properties more simply as the 'traffic,' 'rank,' and 'connectedness' of a page, respectively. Notice that each attribute superficially appears to be determined by only one type of thing: users determine traffic, search engines determine rank, and pages (and by implication authors) determine connectedness. However, in reality, all three properties are deeply intertwined; but it was not always this way.

In the beginning of the Web, there were no search engines, only links. As a result, users could visit a Web page only by directly typing in a URL (the part at the top of your browser that typically begins with 'http://'), or by clicking on a link. Relative to each other, a click is far easier for a user to do than it is to type in a URL. This leads us to the first observation on the relationship between traffic and connectedness:

The greater a page's connectedness,
the greater its traffic.

After all, if users predominately arrive at pages via a link, then (all things being equal) the more pages that link to a certain page, the more clicks from different locations that it can generate.

Different stages of the Web also saw vastly different demographics between Web page users and Web page authors. Given the Web's academic origin, most early authors were scientists, as were most users. But as excitement for the Web spread, and being that it is far easier to be a user than an author, there was a brief period in time in which users and authors were very different groups of people.

Over time, as Web-authoring tools became readily available and as Web resources became easier to attain, these two demographics gradually merged. Thus, in the current state of the world, many Web users are also Web authors. We will explore this fact more closely later when we discuss the phenomenon of Web loggers. However, for now, just consider the fact that when authors and users come from similar pools of people, a new relationship emerges:

The greater a page's traffic, the greater its connectedness.

This happens simply because people tend to link to pages that they themselves value.

Still in the dark ages of the Web, there suddenly emerged a new tool: the search engine. Now ubiquitous, the first general purpose search engine, AltaVista, represented a revolution in usability on the Web. Suddenly, pages could be found by content and not just by location. Instead of knowing where some piece of information was located on the Web, one could find it by supplying a rough sketch (say a few keywords) to describe the desired document. While there are many benefits to retrieving information in this reversed manner, there is an unfortunate side affect: a single query can have thousands or even millions of valid results. Worse yet, some results, while technically a valid match to a query, may actually be off topic to the intent of a user's query. For these cases, the 'right' result may be buried deep within a pile of 'wrong' results.

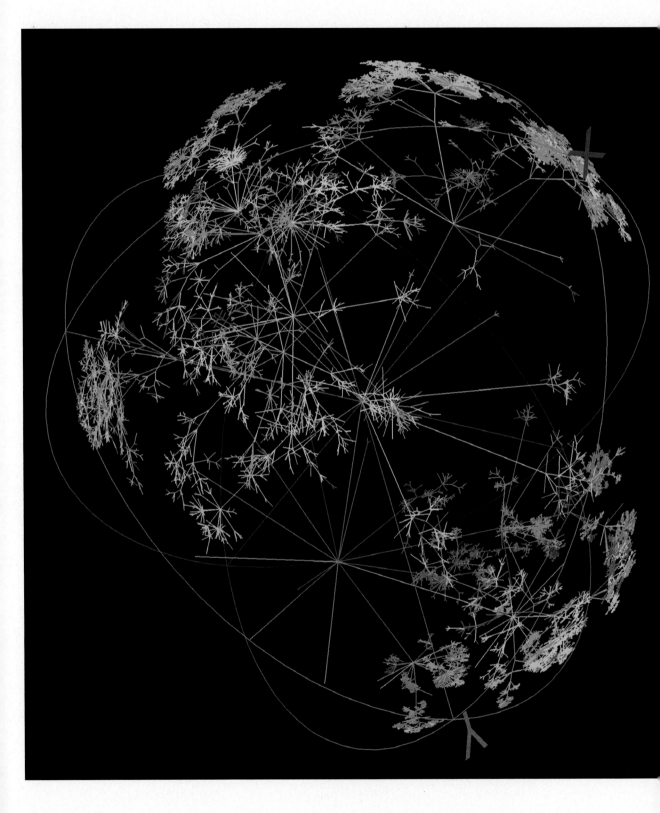

Search engines – back then and still today – try to ease the burden on the user by ordering search results so that the high quality pages that are likely to satisfy a user's intent show up first. But the process of ranking results is both art and science and still far from perfect. In any event, with the emergence of search engines came a new relationship:

The greater a page's rank, the greater its traffic.

This new relationship holds simply because a search engine can introduce users to pages that they never knew about. Moreover, in the case where the user is an author as well, we also find the corollary:

The greater a page's rank, the greater its connectedness.

Hence, the programs behind the search engines have an impact on the traffic patterns of users and the linking patterns of pages (as search engines influence authors).

Over time, new search engines would come and go, offering different features with the goal of earning a dedicated user base. But the sticky feature – a feature that entices users to be repeat users – is a better ranking function, one that seems to anticipate the user intentions, and satisfies the user needs with relevant results better than the competition.

Two interesting breakthroughs in the search engine industry used an implicit form of intelligence embedded within the Web: traffic and connectedness. In the aggregate, traffic patterns on the Web reflect what users find valuable, while patterns in connectedness reflect what authors find valuable. Both represent something akin to a voting scheme for ordering pages by value. In the late 1990s each of these ideas were exploited by two new search engines, DirectHit and Google, which were able to use traffic and connectedness (respectively) to more effectively rank pages.

Today, virtually every major search engine uses traffic and connectedness as an ingredient to their ranking function, but at the time of their introduction, DirectHit and Google each represented another major step in search engine technology by using the collective wisdom of the Web to better satisfy users. However, these two innovations closed the loop, so to speak, on how people, programs, and pages influence one another:

The greater a page's traffic, the greater its rank.
The greater a page's connectedness, the greater its rank.

With these final two relationships, people, programs, and pages each have the ability to influence one another. We have seen throughout this book how circular relationships (i.e. positive feedback loops) are key to the creation of fractals and chaos, and so it is on the Web. Besides the benefits seen from an evolving Web, we can also see instances of spontaneous weirdness that are all a direct consequence of the Web's recursion:

• A single link from an influential Web site can cause the linked Web site to collapse, due to a spontaneous increase in traffic. For example, the Web site Slashdot, http://slashdot.org/, is a daily compendium of links to interesting developments in technology, submitted by a vast and sometimes fanatical user base, and vetted by editors. When Slashdot adds a new link to an interesting Web page, the ensuing stampede of readers clicking on the link can bring an unprepared Web site to its knees under the weight of all its new audience.

We have seen throughout this book how circular relationships (i.e. positive feedback loops) are key to the creation of fractals and chaos, and so it is on the Web.

This phenomenon has been called the Slashdot effect (even if the originating site is not Slashdot itself), and affected sites are said to be slashdotted.

- Communities of Web loggers have colluded to form Google bombs. By collectively linking to a page in an atypical manner, small groups of individuals have successfully tricked search engines into producing humorous results. For example, a search on 'more evil than evil itself' used to return Microsoft's Web site as the top-ranked result. This was accomplished by a loosely coordinated group of Web authors creating links to Microsoft, where the underlined text in the link (the so-called anchor text) said 'more evil than evil itself'. A similar phenomenon is known as link spam where individuals attempt to influence search engines to favor pages of their choosing.

All of these cases are a consequence of the Web reflecting an intricate coupling between people, programs, and pages. Throughout the rest of the chapter we will see how the Web's recursion yields a surprising degree of self-organization, self-regulation, and self-similarity on multiple levels.

The Macroscopic Web

Assigning superlatives to the Web is easy: it's massive, it's dynamic, it's decentralized – it's unlike anything else in the world. But one of the Web's most amazing attributes is that it is arguably the largest self-organized artifact in existence. Every day millions of Web publishers add, delete, move, and change their pages and links, yet what results is far from random or haphazard. Rather, from these millions of uncoordinated decisions emerge a startling number of regularities. Figures 6.1 and 6.2 display two visualizations of the Internet's map, its complex flowering and branching structures tantalizingly fractal-like. Scientists have quantified that

intuition, uncovering self-organizing fractal patterns in examining nearly every aspect of the Web, including the contents of pages, the hyperlinks between pages [Barabási 1999], the physical wires making up the Internet [Faloutsos 1999], the types of files found on the Web [Crovella 1998], the traffic patterns on the Internet [Leland 1993] [Crovella 1996], and the behaviour of people as they surf the Web [Huberman 1998].

Consider traffic patterns. If you were to tap a particular wire on the Internet and listen as emails, Web page contents, and other data zipped back and forth, you would observe erratic rises and falls in the volume of traffic, marked with occasional bursts. Figure 6.3a shows a representative sample of traffic volume over the course of 100,000 seconds, or a little more than a day: you can see somewhat noisy fluctuations punctuated with large bursts. Figure 6.3b zooms in on a particular 10,000-second sub-period (about three hours) within the full series. The pattern of fluctuations and bursts looks roughly the same. Similarly, in Figures 6.3c through 6.3e, as we zoom in to shorter and shorter time scales, the same degree of fluctuations and bursts seems evident. The distribution of traffic is neither smoothing out nor getting choppier as we zoom in further and further. Here we have the classic appearance of self-similarity. We observe the same statistical behavior regardless of the resolution (time scale) of our plot. Scientific studies confirm mathematically what our eye suspects: statistical measurements of the variability of traffic on the Internet and on corporate networks do not differ substantially whether we are examining patterns across a month, a day, an hour, or a few seconds [Leland 1993] [Crovella 1996].

Why is Internet traffic self-similar? The answer is surprisingly simple. A particular wire on the Internet will carry a variety of data traffic, including email, Web pages, images, music, videos, and network control information. Each piece of data requires a different

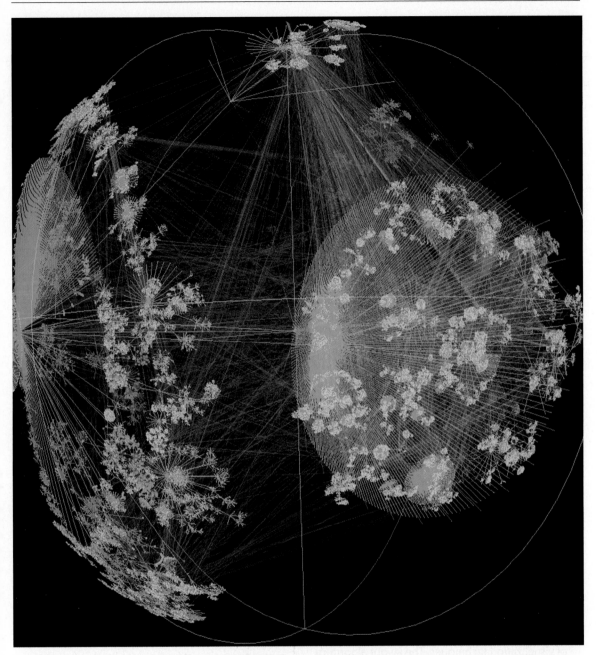

amount of information to encode: a single email usually requires little information, while a video clip of a movie trailer requires much more information. While the vast majority of pieces of data travelling around the Internet are quite small, a few pieces of data are many many times larger than average. The occasional video file or dissertation-length email punctuates a steadier stream of comparatively miniscule Web pages, emails,

Fig. 6.2 above and previous page: Two additional visualizations of part of the Internet's topology, each generated after a day of probing the Internet from a single source. The topological structure is rendered inside a sphere using hyperbolic geometry, which yields a fisheye-like display.

etc. This skewed distribution in the sizes of pieces of data is called a power law distribution or a heavy-tailed distribution, for reasons we will explain shortly. It turns out that, when the sizes of pieces of data in a stream of traffic are governed by a power law, that stream will be self-similar. That's all it takes for self-similarity to arise. The consistency of the series of plots in Figures 6.3a through 6.3e is a direct result of the fact that data traversing across the Internet is mainly a river of small and moderate bits of data littered intermittently with relatively monstrous chunks.

In more detail, a power law states that, within a set of items, items of size x are a constant factor (say, two times) more frequent than items of size 2x. In turn, items of size 2x are twice as frequent as items of size 4x. Mathematically, the frequency of an item of size x is proportional to $x^{-\beta}$, where β is a constant. For example, suppose $\beta=1$. Then the frequency of an item of size 2 is 2^{-1} or $\frac{1}{2}$, while the frequency of an item of size 4 is 4^{-1} or $\frac{1}{4}$. The larger the item, the less frequently it occurs, in direct proportion to its size.

Visualizing the Net

Creating a visual depiction of the Internet is no easy task. The difficulty is not only a matter of the Internet's size. Because the Internet is composed of independent computers distributed around the globe, no one person can hope to compile a specification of all the computers and connections involved. Visualization is also hampered by the fact that the overlapping connections in the Internet – and similarly the hyperlinks among Web pages – are impossible to flatten into two-dimensional or three-dimensional images suitable for human consumption.

Scientists have long examined the problem of visualizing high-dimensional data in two or three dimensions. Throughout this chapter, we report summary characterizations of statistical measures of the Internet that we can show using traditional two-dimensional plots. Figures 6.1 and 6.2 represent more direct attempts at capturing the structure of the Internet in images, using a variety of visualization techniques. The layout algorithm used for Figure 6.1 can take almost a day of computing time to optimize visual space. The method used for Figure 6.2 was developed by Young Hyun, and was based on the pioneering visualization techniques of Tamara Munzer. By plotting points within a three-dimensional sphere, the image is more comprehensible for viewers and allows a natural interactive mode where different points can be 'dragged' into the centre of the sphere for closer inspection of that point and its neighbourhood.
(http://www.caida.org/tools/visualization/walrus/)

A number of other scientific efforts have focused on depicting the intricacies of the Internet using visual means. Many are cataloged in The Atlas of Cyberspace [Dodge 2002].
(http://www.cybergeography.org/atlas/) Ben Fry of the MIT Media Lab has created a real-time animation of Web traffic, growing and squirming like an anemone in immediate response to browsing behaviour across an MIT Web site (http://acg. media.mit.edu/people/fry/anemone/).Beyond mapping, several teams have explored methods for presenting Web search results graphically, though none has yet supplanted today's standard text-based lists.

To many people, the inner workings of the Internet are a mystery: how do computers everywhere interact so that email and Web contents zip to and from the right places at the right times? An informative and entertaining computer-animated movie called The Warriors of the Net (http://www.warriorsofthe.net/) explains the Internet's mechanics by portraying its components (bits, wires, packets, routers, firewalls, etc.) as robotic creatures in a stark factory of the future.

Fig. 6.3 below: Self-similarity of Internet traffic. Fluctuations and bursts in traffic over a period of (a) 100,000 seconds, or about one day (b) 10,000 seconds, or about three hours, (c) 1,000 seconds, (d) 100 seconds, and (e) 10 seconds. Each plot is a zoomed-in image of the previous. The degree of fluctuations and bursts appears similar at every level.

Figure courtesy Will E. Leland et al., 'On the Self-Similar Nature of Ethernet Traffic,' ACM SIGComm'93, p. 186, Copyright © 1993 ACM, Inc. Reprinted by permission.

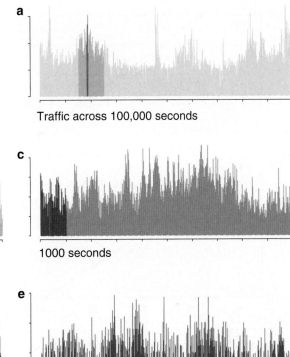

Traffic across 100,000 seconds

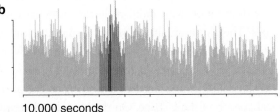

10,000 seconds

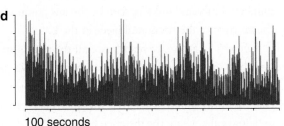

1000 seconds

100 seconds

10 seconds

The distribution is called a power law because of the constant power β used in the formula for frequency.

The distribution is said to be heavy-tailed because the tail of the distribution (the right-hand side of the distribution when plotted, or the part describing large values of x) actually contains a much larger proportion of items than would be predicted by the standard *bell-shaped* distribution (a.k.a. the Normal or Gaussian distribution) often used in statistics. That is, as we move to the far right of a bell-shaped distribution – well past the centre of the bell – the frequency of items approaches zero extremely quickly, much more quickly than in a power law distribution.

The power law is a fundamental indicator of fractal-ness [Schroeder 1995]. A power law is such that, no matter how much we zoom in or out, it looks the same. It doesn't matter if we draw a plot of the distribution over a huge range of sizes, say ranging from 1 to 100,000, or over a smaller range of sizes, say between 10 and 100, the shape of the distribution will be the same. Figure 6.4 illustrates the self-similar nature of the power law.

Power Laws and the Log-Log Plot

The best way to understand the power law is by example. In Figure 6.5, we show a series of plots, all displaying the same information in different ways. All

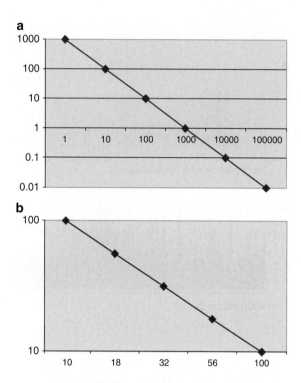

of the plots convey information about the number of inbound links to each of 100,000 randomly chosen Web pages. Each point on the graph can be read as follows: the point's x-value is a particular number of inbound links, while the point's y-value is the number of Web pages (among the 100,000) that have the specified number of inbound links pointing to them. This type of plot, which displays the number of items that appear within specified ranges on the x-axis, is called a distribution or a histogram. Figure 6.5a shows the distribution with ordinary linear scales on each axis. The plot is an almost perfect L shape, revealing the extremely skewed distribution of links on the Web. Almost all Web pages have a very small number of inbound links, as seen by the points lying on the vertical portion of the L shape. On the other hand, a tiny handful of Web pages have a hugely disproportionate number of inbound links, as seen by the few points on the far right of the horizontal piece of the L.

Figure 6.5a is hard to read, since all the points are squashed onto the vertical and horizontal pieces of the L. Figure 6.5b displays exactly the same information: the only difference is that the x-axis is plotted on a log scale, where the distance between the x-values of one and ten is given as much visual space as the distance between ten and one hundred and the

Fig. 6.4 above: Self-similarity of the power law distribution. Both plots show the same power law distribution with parameter β=1, so that frequency equals $x-1$. The top graph displays a large region from 1 to 100,000; the bottom graph displays a smaller region from 10 to 100. No matter what region is plotted at what resolution, the distribution will always appear as straight line (of the same slope) on a log-log plot.

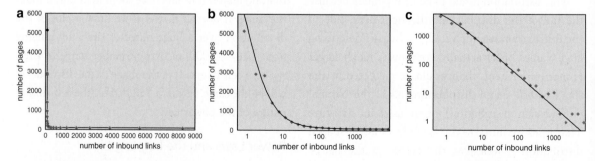

Fig. 6.5 Different ways to visualize a power law distribution. All three graphs display the same data: a histogram of the number of Web pages (among a random subset of 100,000 pages) that have a specific number of inbound links pointing to them. (a) Shown with both axes on a linear scale; (b) The horizontal axis with a logarithimic scale; (c) both axes on a logarithmic scale.

distance between one hundred and one thousand. The log scale stretches out the data points, making it easier to see the detail of the distribution.

Figure 6.5c plots the same information using log scales on both the x and y axes. This is the so-called log-log plot. Notice that, once the data is drawn on a log-log plot, a striking regularity emerges that would be impossible to see using the linear scales of Figure 6.5a: the points follow an almost perfectly straight line. When a distribution drawn on a log-log plot follows a straight line, it is a power law distribution.

Power laws arise naturally. The amount of wealth spread among people follows a power law. The number of people spread across cities follows a power law. The number of connections in the metabolic network of a microorganism, the number of citations to academic papers, the number of connections in the electricity power grid, and the number of people

seeing a particular movie are but a few of thousands of examples of naturally-occurring power laws.

Power laws also abound on the Web. As mentioned, the sizes of data pieces as they flow across the Internet are distributed according to a power law. The sizes of files themselves, residing on Web servers on the Internet, obey a power law. The number of queries submitted to search engines, the frequency of word usage on pages around the Web, the number of hyperlinks pointing to and from Web pages, the depth to which Web users surf, and the number of physical wires connecting to Internet hubs all follow power laws.

Let's examine more closely the pattern and formation of links on the Web. Figure 6.6a shows the distribution of inbound links on the Web plotted on a log-log plot. Notice that on a log-log plot a power law distribution appears as a straight line. We see that the distribution of inbound links on the

Fig. 6.6 right: A tour of the power-law Web. Distributions capturing nearly all aspects of the Web follow a power law, including (a) inbound links, (b) outbound links, (c) files sizes, and (d) the physical Internet itself (the wires connecting computers around the world). Power laws crop up elsewhere too, including people's behaviour as they surf the Web, and even the level of interest among advertisers to be showcased in conjunction with particular search queries.

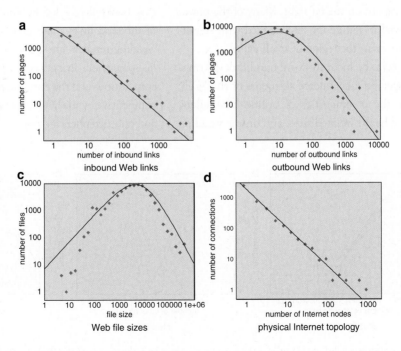

Web is close to a pure power law, except for a very slight drop-off from a straight line at the top left of Figure 6.6a (the region of small values of x, or small numbers of inbound links).

Compare Figure 6.6a with Figure 6.6b. The latter shows the distribution of *outbound* links on the Web (links emanating *from* Web pages) instead of inbound links. Near the right end of the graph, the distribution looks very much the same as the inbound link graph: a straight line on a log-log plot. But on the left side, in the range of smaller values of x, the distribution deviates fairly strongly from the linear signature of a power law. There is a bump in the distribution of outbound links not seen in the graph of inbound links. It turns out that bumps like these are the rule rather than the exception (in this sense, the near perfectly straight line of Web inbound links is rare). For example, the graph of the distribution of file sizes pictured in Figure 6.6c has an even more pronounced bump before straightening out on the far right. Many of the power laws observed in nature are also marked with significant deviations in the region of small values of x.

Figures 6.7a, b, and c show inbound link distributions for specific e-commerce segments of the Web, comparing the communities of online booksellers, commercial health-related sites, and online wedding retailers, respectively. Here we see more examples of the modified power law: in each case, the plot displays a significant bump on the left side before converging toward the linear power law on the right-hand portion of the graph.

In the section that follows on the microscopic web, we will examine what low-level forces are at work in generating both the pure power law seen for inbound links and the modified 'bumpy' power law more common in other distributions. For now, simply note that the closer a community's distribution is to a linear power law, the more cutthroat the competition is to get noticed within that community, and the harder it is for new entrants to compete with the well-established players. The larger the bump on the left edge of the graph (the larger the divergence from a pure power law), the more egalitarian is the community, and the easier it is for new sites to rise to (or near) the top. From analyzing the data underlying Figures 6.7a, b, and c, one can infer that booksellers – led by Amazon.com with millions of inbound links – are extremely competitive, while wedding retailers are less so. Commercial health sites lie somewhere in between. Similarly, online sites for corporations and the entertainment industry are highly competitive, while Web sites for scientists, universities, and photographers are not.

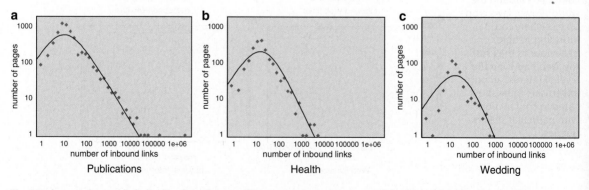

Fig. 6.7 above: Inbound link distributions for specific e-commerce segments of the Web. (a) Online booksellers, (b) commercial health-related sites, and (c) online wedding retailers. Each plot shows increasing divergence from a pure power law, indicating decreasing competitiveness within that community.

There are multiple factors that can lead to the differences in competition that we see. For commercial wedding sites, one factor could be their local nature: many wedding-related retailers serve only a local area, and those serving different areas usually do not compete. Another factor may be that people looking for wedding services use methods other than the Web more often (e.g. referrals from friends). Perhaps because people use wedding providers rarely, they are less likely to create and share information among related sites on the Web.

Note that more difficulty competing with existing popular sites does not mean that substantially better newcomers cannot become popular quickly. For example, Google (a relative latecomer to the search business) has captured a huge fraction of the Web search business largely by providing better service and spreading through word of mouth.

The Web is a Bow Tie

In 2000, a collaboration of scientists from AltaVista, IBM, and Compaq [Broder 2000] discovered a fascinating property of the Web: somehow, all of the billions of pages and links have organized themselves into an overall bow tie shape as pictured in Figure 6.8. The centre of the bow tie is a core of strongly connected pages: every one of these pages can be reached from any other page within the core by clicking on a sequence of links (the sequence may need to traverse a number of intermediate pages, but some path exists between the two core pages). The left bow is connected to the core, but only through outgoing links. That is, there exist links from the left bow to the core, but not vice versa. Conversely, the right bow is connected from the core only via inbound links. One can traverse links from the core to the right bow, but not back again. Finally, disconnected pages that have no links either to or from the core surround the bow tie. The scientists measured the relative sizes of these four main components of

Zipf's Law

The Web is not alone in exhibiting power laws. Data gathered on the use of language, the population of cities, and the distribution of wealth all show clear power-law behaviour. The frequency of words used in human language deserves special mention: it follows the famous Zipf's Law, named after George Kingsley Zipf, an early twentieth-century scientist who revolutionized our understanding of power laws, and helped to reveal their astonishing prevalence throughout society and nature.

Zipf's Law states that the most common word used in language is a constant factor (say, two times) more common than the second most common word, and the second most common is twice as common as the third, etc. Remarkably, for almost any sizeable source of words you can think of – all *New York Times* articles, or all the works of Shakespeare, or all textbooks on molecular biology, or the Bible – Zipf's law holds.

In 1955, Herbert Simon sought to unify the observations of Zipf and others by formulating a single common explanatory model for many of the systems displaying power-law behaviour, including language, population, and wealth. Benoît Mandelbrot [1953,1959] proposed a fascinating alternative explanation for Zipf's Law as it relates to language. He showed that the distribution can be understood as the end result of centuries of adaptive maximization of the information content of language.

the Web (the core, the left bow, the right bow, and the disconnected pages). To their surprise, all four components were roughly the same size.

A year later, some of the same scientists [Dill 2001] showed that the bow tie property is a feature not only of the Web in its entirety, but also of various pieces of the Web. No matter how the Web is sliced – whether by content into topic-specific clusters, by geographic location into regions, or by

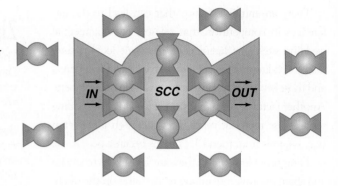

Fig. 6.8 The Bow Tie structure of the Web, consisting of a strongly connected component (SCC), a set of pages that follow into the SCC, a set of pages that pass out of the SCC, and a set of smaller disconnected islands that are themselves mini bow ties.

organizational entity into groups of pages owned by the same person – the bow tie shape emerges, even retaining the rough equality of size among the four main components of the bow tie. This strange structure appears endemic to the Web and pervasive at all levels, revealing a beautiful new type of self-similarity not seen anywhere else.

Search Engines: Tapping The Ebb and Flow of Ideas

In a way, search engines like Google and Yahoo! have a window into the mind of the masses. Search queries stream in by the second capturing people's thoughts, worries, and whims, whatever they happen to be looking for at that particular time. Web sites like Google's Zeitgeist:

(http://www.google.com/press/zeitgeist.html),
the Yahoo! Buzz Index (http://buzz.yahoo.com/),
and the Lycos 50 (http://50.lycos.com/)

report on fads and trends reflected in search traffic: the thoughts and ideas that people are searching for en masse, including what is hot and what is passé. It is fascinating to watch as memes appear, skyrocket, cycle, or decay, as the case may be.

In watching the top search terms from one week to the next, clearly some terms will stay perched among the top ten, while others will drop out. For example,

as of Sunday August 3, 2003, 'Britney Spears' moved from third to second place, continuing a remarkable run of 123 straight weeks atop the Yahoo! Buzz Index charts. 'Tour de France' also remained in the top ten, though only for the second week running. Meanwhile, 'Beyoncé Knowles' and 'PlayStation 2' fell from their top-ten perch the prior week, supplanted by Kobe Bryant and Angelina Jolie, celebrities whose profiles rose during the week, fuelled by a criminal indictment and a new movie release, respectively. The percent of terms that disappear from the top ten from one week to the next – equivalent to the percent of new terms, and reciprocal to the percent of stationary terms – is called the churn rate. The churn rate of search terms captures the speed at which new memes rise and old memes fall.

Churn rate can be computed for different numbers of top N terms. We can examine the proportion of terms lost from the top ten, or the proportion lost from the top 100, or the proportion lost from the top 50,000 terms. Note also that we can compute churn rate over any time frame: daily, weekly, monthly, etc.

You might hypothesize that churn rates would differ depending on whether you examine the top ten terms, or, say, the top 50,000 terms. For example, it seems reasonable that the status of the most popular terms could be so self-reinforcing as to render them

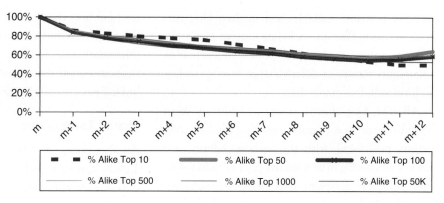

Fig. 6.9 right: Month-to-month churn rates describing how popular search queries shift over time. Churn rates exhibit self-similarity, remaining the same as the number of terms considered ranges from 10 to 50,000.

more stable than the hordes of terms among the top 50,000. However, this hypothesis is not correct: in reality, the top ten is no more stable than the top 50,000. In fact, no matter what value for N is chosen – 10, 50, 100, 500, 1000, or 50,000 – churn rates are unaffected. Figure 6.9 shows churn rates after one month, two months, three months, etc., out to one year. As expected, the longer the time frame, the higher the churn rate, as a greater proportion of terms filter up and down. However, for any given time frame (say, seven months), churn rates are nearly identical for all values of N. Here we see a remarkable form of self-similarity: with no matter what granularity we look at search terms – whether we zoom in to examine the top ten, or zoom out to examine the top 50,000 – the percent of terms entering and leaving the identified set remains constant.

The Middle Web

Having just seen how the Web contains some measure of order at the highest level, we now turn our attention to the next lower level, where groups of authors and users form patterns on the Web. The short version of this story is that the Web's content is effectively self-organized by the actions of individuals. Contrasting this self-organization to the more

familiar phenomenon of centralized organization, we will see that the Web exhibits aggregate behaviour that begins to resemble a hive-like intelligence.

Web Logs a.k.a. Blogs

One of the more recent additions to the Web site bestiary is the Web log or blog. Blogs began as something like online diaries with authors making regular postings that were topically focused on everything under the sun or nothing in particular. Journalists and pundits found the medium to be promising new ground for self-publishing. At its best, early blogs allowed for grass-roots journalism and an unbiased flow of ideas and information. At its worst, blogs were simply vanity sites.

The emergence of blogs is important for two reasons. First, blogs, more than any other phenomenon, blurred the line between author and user as most blog content was about the first hand experience of visiting other Web sites. Second, blog software – the programs that facilitate and automate the maintenance of a blog site – would evolve in sophistication, incorporating many new features including user accounts, discussions, postings by multiple individuals, rating systems (of users and posted stories), multimedia, and search. Today, sophisticated blog software is freely available, and modern blog sites come in many flavours including current event discussions, various grades of self-published journalism, community forums of differing degrees of speciality, and, yet still, the simple diary.

All told, blog sites represent a deliberate effort by individuals to cooperate towards a form of community publishing, with the authors, editors, and readers all coming from a similar pool of individuals. Blog sites also represent larger-scale communities, beyond a single site, because many individuals often contribute to the content of modern blog sites and the membership of related blog sites often overlap. Moreover, the content on one blog site often influences the content on other blogs.

Modern search engines, which use link structure for improving the relevance of served results, have had to co-evolve with the emergence of blogs for multiple reasons. The primary reason is that blogs, by and large, are quirky sites, yet they carry a disproportionate amount of influence in assessing the importance of Web sites because they contain so many links. When a quirky group of people link to pages in an atypical manner, their quirks are propagated to the mainstream if left unchecked.

This amplification property of blogs results in many interesting social phenomena on the Web that has no real-world analogue. Propagation of memes on the Web can start with a single blog site distributing a funny or unusual link. Other blog sites, exhibiting almost a flocking behavior, redistribute the meme, which impacts not only the content that people read but also the links that persist on the Web. In this way, ideas and information (both true, false, and otherwise) can circumnavigate the globe multiple times in a single day, making the circular influence of linking patterns all the more pronounced.

Shared Taxonomies

Another form of deliberate cooperation by Web authors can be found in shared taxonomies, which is best exemplified by the Open Directory Project (ODP) located at http://dmoz.org/. The ODP consists of a topical taxonomy, not unlike the best-known taxonomy at Yahoo!. However, the ODP is a strictly volunteer effort, where individual editors assume ownership for different topics on the Web. The volunteer editors collect links to pages that are relevant to their particular speciality and incorporate them into their respective location within the taxonomy. All told, the ODP has thousands of editors that maintain links to millions of pages, which, in turn, are incorporated into the ranking algorithms of the most important search engines.

Clearly, the ODP is a distributed effort by individuals to bring order to the Web. However, as with blogs, the ODP represents a deliberate and intentional form of cooperation by individuals. There exists an unintentional form of cooperation by authors that is, perhaps, even more striking than the ODP and blogs because it represents the truest form of self-organization; namely, one in which the individuals cooperating do not even know that they are contributing to something larger.

The essence of the self-organized nature of the Web is that authors – being somewhat independent of one another – can effectively do whatever they want.

Hubs and Authorities

The essence of the self-organized nature of the Web is that authors – being somewhat independent of one another – can effectively do whatever they want. Some will post a flat collection of favorite bookmarks about nothing specific. Others contribute volumes of original material that is focused on a single topic. And still other authors produce nothing more than small collections of links that point to things that are all about the same thing. These last two examples of authors – those that create original material and those that point to focused material – are special in that they form two halves of a single relationship.

Web pages that contain compelling original material (without necessarily the emphasis on having many outgoing links) are often referred to as *authority* Web sites, or more simply as just *authorities*. Authorities have the property that they tend to accumulate incoming links because others interested in their content will create links that point to them. The name, *authority*, comes from the language of bibliographic studies where there is a notion of a work of literature as being authoritative if many authors cite it. As with literature, Web authorities are frequently cited but with links instead of proper citations.

A *hub* Web page (or more simply a *hub*) is the complement to the authority. Hubs are akin to a survey papers or focused reference books in that they contain links that point to many pages that are all about the same topic. Hubs are natural organizers of information because they group similar things together.

Together, hubs and authorities form a recursive relationship that reflects the dependencies between the two types of pages [Kleinberg 1999]. While authorities may earn links by having original content, they may also acquire links by the rich-get-richer process alluded to above (and which we will

examine in greater detail in the next section), where highly-linked sites tend to obtain even more links due to their greater visibility. That is to say, hubs may have to link to very popular authorities if they are to retain their status as being a hub. (Not doing so would be like writing a survey article on evolution that fails to cite Darwin.) Similarly, authorities are only truly recognized as being authorities if important hubs link them. Together, these two facts yield a recursive definition for what it means to be a hub or authority.

Hubs are pages that link to authorities.
Authorities are pages that are linked by hubs.

Put simply, these two definitions are recursive because each entity in some sense defines the other. What is truly fascinating about this mutual dependence is that Web pages – in the wild, so to speak – seem to co-evolve via this recursive relationship.

Community Signatures

In 1999, Ravi Kumar and his colleagues [Kumar 1999] surmised that if the Web is, in fact, composed of many hubs and authorities, then one should be able to find a Web community core by looking for a group of hubs that all point to the same set of authorities. Mathematically speaking, these two groups of pages form what is known as a bipartite core. A bipartite structure is illustrated in Figure 6.10, and consists of two types of objects: those in the left set and those in the right set, with every object on the left pointing to every object on the right. Notice that this structure is identical to what you would expect to find if there existed some number of hubs that were all focused on the same collection of authorities.

Kumar et al. found that there were hundreds of thousands of community cores that contained this exact bipartite signature. When inspected by hand, these community cores were almost always focused on

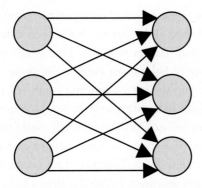

Fig. 6.10 above: A bipartite core on the Web: every page on the left links to every page on the right.

Self-Organized Communities

The link structure of the Web is not unlike the social network of humans. We have reciprocal relationships with some people, and we know of people that don't know us, which are respectively akin to pages that mutually link to each other and pages where one links to the other only in one direction. Who we are is in some sense defined by the links we have in the human social network. Likewise, Web pages can also be better understood by examining the context in which pages exist within a Web community.

The notion of a Web community core, as defined above, is powerful in the sense that it gives an unambiguous signature from which to identify collections of related Web pages. However, this notion can be considered insufficient because most Web pages will not belong to a Web community core. How, then, can one identify the community in which a page belongs?

There are many different ways to define a Web community, and to be sure, there is no absolute correct definition. Nonetheless, some definitions for a Web community can be used to identify largish collections of pages that, in some sense, seem to belong with one another because they are all focused on a similar theme. We now turn to one particular definition for a Web community that is mathematically rigorous in that it is well defined, is surprisingly intuitive and simple to understand, and empirically corresponds well to real communities on the Web.

For reasons to be explained shortly, we will refer to this type of Web community as a cut Web community, or more simply as just a cut community. A cut community consists of a collection of pages that predominately link to one another (with links in either direction). That's the whole definition; it is simple, but yields several elegant properties.

First, note that it is a meta-definition in that it permits one to make more specific statements like 'the bicycle community consists of pages that predominately

an extremely narrow topic such as Japanese elementary schools, Hotels in Costa Rica, or Turkish student associations. But most striking, the identified community cores are often so narrow and specific that they are not contained in any taxonomy like the ODP.

Because Web pages contain both regular content and links, there are multiple ways in which two pages can be said to be similar (or dissimilar) to each other. Ignoring text and focusing just on links, one can easily see that hubs within the same community core have outbound links that are similar or identical. Authorities within the same core have inbound links that are similar or identical. Thus, we can speak of two pages as being similar in content (they express similar words and concepts), in outbound links (they point to approximately the same pages), or in inbound links (they are pointed to by the same inbound links).

One remarkable attribute of the Web is that similarity in inbound or outbound links often implies similarity in page content. This relationship means that one can find new pages of interest by looking only at how pages link to one another within a local neighbourhood of a starting page. The connection between links and content also means that one can analyze link structure to find how topics on the Web relate to one another.

link to bicycle pages'. Also note that community membership is easy to test for and validate. Hence, if you know about the bulk of the bicycle community, you can look at the links coming in and out of a page in question. If more than half of the links refer back to the bicycle community, then the page in question is also a member of the bicycle community.

In 2000, Gary William Flake and colleagues [Flake 2000] discovered an effective method for identifying self-organized collections of Web pages that obeyed the cut community definition. The method works by recasting the community identification problem into what is known as the *s-t minimum cut network problem*. In this framework, one looks at a collection of pages and links and asks: for two pages, s and t, what is the smallest number of links that need to be 'cut' (i.e., removed) in order to completely separate s and t from one another, where s is a page that is indicative of the type of community that one is looking for and t is an artificial page that represents the whole of the Web. By looking for the smallest number of links to cut, the procedure effectively tries to find the smallest group of pages connected to s (our page of interest) that nicely separates from the rest of the Web.

Flake's community algorithm also has the nice attribute that it is computationally efficient. Nonetheless, it is not at all clear that it should even produce collections of pages that are all focused on a single theme. However, in practice, the community algorithm is remarkably successful at finding large collections of related pages. When seeded with the personal home pages of famous scientists, the community algorithm will find hundreds or thousands of pages that are all focused on the specialty of the scientist in question [Flake 2002].

In fact, the community algorithm, and other link-based approaches, have been shown to be very effective in making sense of the Web. Notice the language-independent nature of link-based methods:

since they ignore the textual content of pages, they work equally well for pages in English, Spanish, or Swahili, for that matter, or for pages composed nearly entirely of images and multimedia. But here's a secret of the power of the community algorithm and other methods like it: it's not the algorithm that's special, it's the Web.

Topic Affinity

Consider a completely random Web surfer, who wanders about the Web clicking on randomly chosen links (we will have more to say about the properties and implication of the random surfer model in the next section). The surfer travels from page to page, each time moving forward by clicking on a random link on the current page. Assuming that the surfer starts at a random page, we can measure the relative bond between content and links by measuring how long it takes for the random surfer to visit pages that drift away from the topic of the starting page.

Soumen Chakrabarti and his colleagues [Chakrabarti 2002] found that on the whole, the correspondence between the topicality of a page, and the links that it contains is remarkably strong. In the example of our random surfer, Chakrabarti et al. found that for some subjects, a random surfer could remain on topic after following as many as 5 or 10 links. Interestingly, the degree of topic drift was strongly dependent on the starting topic. For example, 'soccer' pages would drift off-topic relatively fast, while 'photography' pages maintain topical focus for many more steps.

Related to all of this is the role of anchor text to content. Anchor text is the text that is contained in a link (usually underlined in most browsers). The author of a page that contains a link creates the anchor text, but anchor text is usually intended to be descriptive of the page that the link points to, not necessarily the page that contains the link.

Small World Networks

Many people are familiar with the expression 'six degrees of separation' which suggests that for any two people in the world, there are at most six person-to-person relationships that separate those two people. Thus, you and I may not have any friends in common, but we will probably have a friend-of-a-friend-of-a-friend in common.

The remarkable feature of small world networks is that they contain few links relative to their number of members. Intuitively, small world networks have this dual property by having many members with mostly 'local' relationships (say, most of your friends and neighbours), and a very small number of members that have 'global' relationships (e.g., a celebrity that is known or knows thousands of people). Thus, the path that joins any two random people is likely to begin and end with some local relationships, but will pass through some global relationships in the middle.

Throughout this chapter, we have seen how the Web reflects the properties of our society – and so it does with the small world nature of human culture [Watts 1998]. Between any two Web pages in the Web's largest strongly-connected core, there are at most a few dozen links that connect those pages. E-mail and instant messaging relationships also form small world networks.

The good news about small world networks is that for those who know how to pick links to follow, a small number of clicks will lead one to a desirable location. The bad news is that it is also remarkably easy to spread problems (like viruses and misinformation) in a small world network as well.

Computer scientists have long been working on the problem of how to recognize when a document is about a particular topic by analyzing a document's text. In this vein, scientists have used these tools to improve search engines and related technologies. Interestingly, many scientists studying the Web have found that the anchor text that points to a page is often a stronger indicator of the referred page's subject than its own text. This is truly a surprising result because it means that the links that point to a page are often a better descriptor of a page's content than its own title.

All of this goes to show that despite being decentralized, the Web seems to 'like' order instead of disorder. Authors don't have to link to pages that relate to their own; but they do. And authors don't have to use anchor text that is strongly relevant to the pages that it refers to; but they do that too. The bottom line is that links, instead of being unruly, are, in fact, self-organizing. In the aggregate, connections and content go hand-in-hand and they co-contribute to the Web's higher-level formation of patterns and structure.

Having seen the self-similarity evident in the top-level view of the Web, and the self-organizing niches and structures evident in the middle Web, we now turn to the underlying low-level processes and forces driving organization and structure on the Web.

The Microscopic Web

The Web in its most fine-grained detail is the results of billions of individual decisions taken every day around the globe. CNN adds a breaking story; a job hunter updates her online résumé; a university department deletes the homepage of a graduated student; Amazon.com adds a new book title. All around the world, the content of the Web is modified in response to significant real-world events as well as trivial whims. Clearly, to understand the evolution of the Web, we must understand how individual pages are created and modified. However, it is simply impossible to factor into our understanding the details of the innumerable human motivations behind all pages in the Web.

Instead of focusing on minutiae, we need to abstract away as many inessential details as possible and look instead for the simplest rules that capture the most important aspects of the Web's behaviour. This modelling process will help us identify what are the essential ingredients responsible for the self-similar structures on the Web. But don't be fooled by the simplicity of the models that we will talk about. Despite their simplicity, they capture a considerable amount of the complexity that we've observed so far. As Ian Stewart eloquently states elsewhere in this book, simple explanations for complex observations lie at the heart of modelling nature, and fractals are a powerful tool in the mathematician's arsenal for doing so.

Modelling Web Growth

As in the previous section, let's temporarily ignore Web page content and focus on the links that they contain. Moreover, let's also ignore the direction of all links and just focus on the fact that two pages can be linked to one another. To better understand how the Web evolves, we need to understand how individual pages contribute to the overall link structure. We will model the Web's evolution by iterating over the following five steps:

1. Create a new page, called p.
2. Randomly pick an existing link, l, not connected to p.
3. Randomly select one of l's two adjacent pages, q.
4. Add a new link between p to q.
5. Repeat steps 2–4 a total of k times.

One pass through these steps adds one new page to the Web and k new links. Obviously, one can repeat the entire process multiple times to add many new pages and even more links.

The recipe above specifies what is known as a *generative model* because it explicitly shows how one takes a current snapshot of the Web, and generates a successor to it that has grown a little bit. The model – introduced for the Web by Albert-László Barabási and Reka Albert [Barabási 1999] – is simple enough that with sufficient mathematical tools, one can effectively see how it would behave if iterated for an infinite number of steps.

The most interesting part of the recipe for growing the Web is in steps 2 and 3, where we pick a random page, q, in which to connect to our new page, p. If there were n existing pages, and we were to select one of them purely at random, then we would find that each page has a $1/n$ chance of being selected. But that's not we are doing. Notice that we are picking a link, and then picking one of the two pages adjacent to it. This means that the more connected a page is, the more likely it is to be selected in steps 2 and 3.

The selection process in steps 2 and 3 can be reasoned as follows. Think of each link as owning two lottery tickets, and giving away one ticket each to the two pages connected to that link. Now you can verify that the probability that an existing page is selected in steps 2 and 3 is equal exactly to the number of lottery tickets it has, divided by the total number of lottery tickets possessed by all pages. Thus, the more links (or lottery tickets) a page has, the more likely it is to 'win' by being selected in steps 2 and 3. If a page has many links, it's bound to get more. But if a page has few links, it will probably not get many more. As a result of these facts, this pattern is often referred to as a 'rich get richer' phenomenon or as 'preferential linking'.

Clearly, Web page authors don't add links to their pages in precisely this manner. But, as we have seen, more links to a page implies many things, including more traffic, higher ranking, and more visibility, in general. Thus, it is not too outrageous to simplify things and just simply say that authors prefer to link to pages that are more connected.

While some may debate the fairness or desirability of such a state of affairs, the 'rich get richer' process is a common one that arises naturally in a large number of domains, including many social, biological, and physical systems ranging from the power grid network to the metabolic networks of microorganisms.

In an influential article published in the journal *Science* in 1999, Barabási and Albert revealed a fascinating discovery: their simple generative model for Web growth is sufficient to replicate many of the key features of the Web. Most notably, the structures generated using this simple model exhibit precisely that same power law distribution as observed on the real Web, and as we witnessed earlier in the section on the macroscopic Web.

Power Laws and Communities

Barabási and Albert's first Web model touched off a wave of research aimed at capturing additional aspects of the real Web. While Barabási and Albert's model succeeded in capturing some of the highest-level properties of the Web (as well as showing that the Web is unambiguously fractal in construction), it was somewhat incomplete in that it did not account for how the Web operates when viewed on intermediate scales.

David Pennock and his colleagues [Pennock 2002] made a simple modification to the Barabási-Albert model that would account for some of the behaviours of Web communities. Before we get into the details, let's recap some of the intuition behind power laws and how they occur in nature.

Within the biosphere, we see far more small creatures than we do larger creatures: there are many more

bacteria than there are insects; there are many more insects than medium-sized animals; and there are still far fewer large animals, such as whales and elephants, than just about anything else. The distribution of sizes of creatures across all species is a power law.

On the other hand, within a species, we see a different pattern entirely. The size, weight, and height distributions of humans follow the more familiar bell-shaped or *Gaussian* distribution. This means that most individuals fall somewhere in the middle – that is, there are more average-sized people than small people or large people. The trend of having more average individuals than big or small is found in just about all species, when a species is examined in isolation.

Returning to the Web, and thinking about Web communities and inbound links as being somewhat analogous to species and the size of animals, we find that if one looks at the number of inbound links to a Web page – but restricted to pages in the same community – the distribution is neither a strict power law, nor is it Gaussian. Instead, it is bump-shaped (like a Gaussian distribution) but on a logarithmic scale (like a power law), as we saw in Figure 6.7. The important point in all of this is that at the intermediate level, there is something different going on than the strict rich-get-richer linking patterns that the Barabási-Albert model suggests.

As a people, we all know of celebrities such as famous actors, athletes, and politicians. But we also know many people based on our interests, where we live, and where we work. Likewise, Web authors create links not just to popular pages, but also to pages that are related to their own page in some manner. These non-popular links are akin to the people that we personally know, while the popular links (say to Yahoo!) simply represent an awareness of what the masses link to or know of in the aggregate (like a celebrity).

In an article published in the *Proceedings of the National Academy of Sciences* in 2002, David Pennock

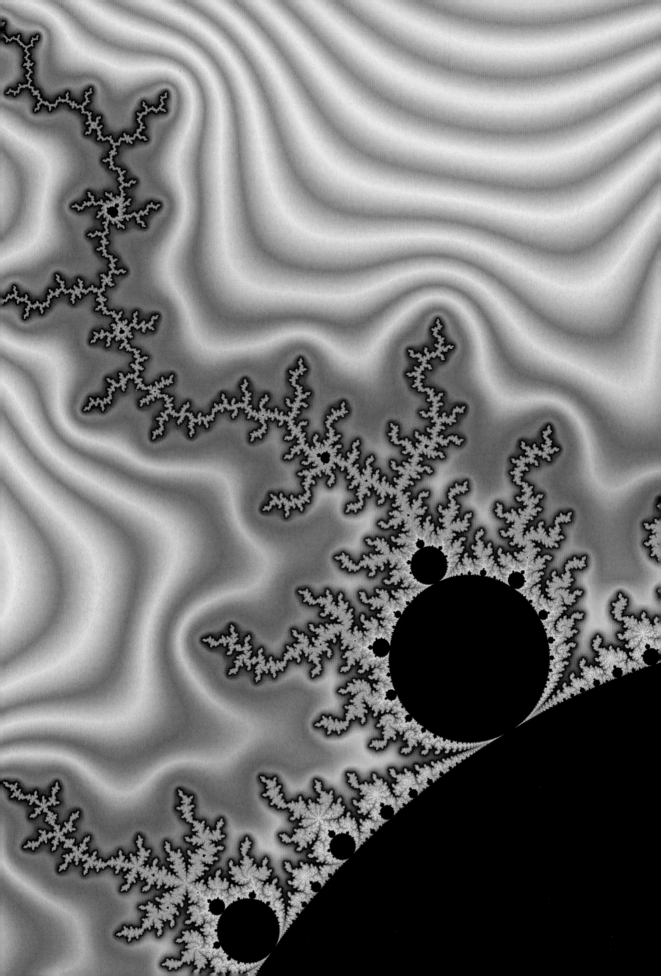

Generative Fractals

The Web (and the output of the Barabási and Albert's model) may not look like a fractal to the casual observer, partly because it does not lend itself to visualization the way other fractals do. Nonetheless, the Web is just as much a fractal as the more familiar eye-pleasing fractals. It is just a little too much for the human eye to behold. However, we can see similarities between the Web and other fractals when we examine how each is produced.

L-systems, discovered by Aristid Lindenmayer [Lindenmayer 1968], simulate plant growth with only a small number of rules that specify how 'cells' grow into other cells. As can be seen, each iteration looks like how one would expect a plant to grow. Different 'seeds' and growth rules can be used to produce different types of plant-like structures.

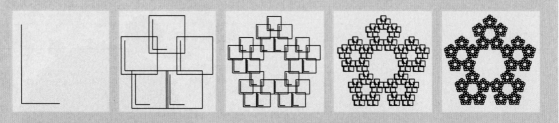

MRCM fractals are produced by iteratively expanding parts of the fractal so that each part contains a smaller version of the whole. After a few iterations, the MRCM fractal will possess the signature look and feel of a fractal.

In both cases – as well as in the Barabási and Albert Web model – taking one stage, applying a simple rule to it, yields the next stage, and ultimately produces a fractal

and his colleagues showed two important things. First, they showed that the distribution of inbound links for category-specific subsets of the Web, for example all University homepages or all movie homepages, follows the power law but bump-like pattern seen in Figure 6.7. Second, they showed that a simple modification to the Barabási-Albert model predicts the observed data on the Web with remarkable accuracy. The new recipe looks like the following:

1. Create a new page, called p.
2. Flip a bias coin; if heads:
   2a. Randomly pick an existing link, l, not connected to p.
   2b. Randomly select one of l's two adjacent pages, q.

3. Else, if tails

    3a. Randomly pick an existing page, q, not equal to p.

4. Add a new link between p to q.

5. Repeat steps 2-4 a total of k times.

The only new difference is that in step 2, we randomly switch between two types of new links, one that is preferentially based as before (in steps 2a and 2b) and one that is uniformly selected in 3a. In step 3a, the page we pick, q, is independent of how many links it has. The 'biasness' of the coin in step 2 influences to what degree the links tend to be preferentially based or non-preferentially based.

Looking back at our analogy between people, step 3a is akin having an association with a person that is not influenced by popularity, which is the main force behind the observed bump in the community link distributions. It turns out that this divergence from the pure power law, while most pronounced within topically coherent communities on the Web, also shows up to a lesser extent in a wide variety of distributions, including the distribution of *outbound* links on the Web and the distribution of movie actor collaborations.

Web Surfing Patterns

Ultimately the Web is about people. Above our focus was on the behavioural patterns of Web authors. In this section, we focus on users. How do people typically surf the Web? Again, we won't get far by analyzing the intricacies of each and every surfer during each of their Web use sessions. Instead, we look instead for overall patterns of behaviour and simplified rules of thumb that seem to capture the essence of observed aggregate behaviour.

Along these lines, Bernardo Huberman and his colleagues [Huberman 1998] developed a simple and elegant model of surfing behaviour. In their model, a user continues to click deeper and deeper

on a particular path of linked pages until he or she reaches a page of sufficiently low perceived quality; at which point the user abandons the current path and either gives up or begins anew, for example by typing in a URL directly, choosing a favorite bookmark, or initiating a web search.

Huberman and his colleagues showed that – assuming surfers on the whole obey the above tendencies – the depth to which the typical user surfs follows a type of power-law distribution called the inverse Gaussian distribution. In fact, data gathered from several different websites and user bases, over different time periods, match the conclusions of the model extremely well. Webmasters can even use the model to predict which pages will receive the most traffic on their site, and how to rearrange their site to maximize traffic to particular pages. Hence, users, in the aggregate, seem to surf Web pages in a fractal manner.

Another model of surfing behavior is called the 'random walk' model, and aptly so. Imagine a completely random surfer. Starting at a random page, this wandering surfer clicks on a random outgoing link, bringing him or her to a new page. From there, the surfer clicks another random link, moving to a third page, etc. The surfer continues like this ad infinitum, except that occasionally (with some small probability at each step) the surfer restarts, 'teleporting' from its current location to a completely random location. [1] Although the random walker model is by any measure an extreme simplification of reality, it turns out to be very powerful.

Because of the teleportation step, we know that the random walker can always move on to a new page. The key question is: which Web pages will the random walker visit most often if allowed to walk forever? It turns out that this question can be answered very elegantly with a remarkably straightforward calculation. The equations behind the calculation are very simple, but they must be performed for every page on the Web multiple times. Instead of diving into those mathematical details, we will instead try to capture the intuition of the random walker, which gets at the heart of what it means for a page to be important. If you think of a link as being an endorsement by an author that the page at the other end is high quality, then we get the following recursive rule:

Web pages are important if other important pages predominately link to them.

In the above rule, we use 'predominately' to mean that the page with the link has a relatively small number of outgoing links and, hence, is only 'voting' for a small number of other pages. (More outgoing links can be interpreted as an author diluting his or her vote.)

Larry Page and Sergey Brin, the founders of Google, discovered in 1998 that this calculation – which they dubbed PageRank after Larry Page – was very effective at separating quality pages from poor pages [Brin 1998]. In fact, when introduced, Google, with the help of PageRank, offered such a vastly better way of organizing the Web that Google came to lead the Web search industry.

The power of PageRank is that it uses the links of the Web (which are made by authors) and simulates how an infinite number of users given infinite time would visit those pages. The pages visited the most by the random walkers are deemed the best. Hence, Google makes explicit use of Web authors and implicit use of users to do a better job of finding quality content.

The Web as a Mirror

We've now come full circle. Having examined the Web from a variety of scales and viewpoints, we have now seen how users, authors, and search engines all influence one another to yield an amazing array of self-organization, self-regulation, and self-similarity.

Ultimately, the Web's organization is intimately related to the complexity of human culture and to the human mind, and it is this subtle relationship between humanity and the Web that is responsible for the Web's amazing properties. In the remainder of this chapter, we will explore how the Web can be seen as a mirror to humanity, and we make some predictions as to where the Web is evolving.

Search and the No Free Lunch Theorem

What computer scientists refer to as 'search' is perhaps the hardest mathematical problem in existence. By 'search' computer scientists mean all of the following:

- Teach a computer to drive with only positive and negative reinforcement. That is, reward the computer when it gets to a destination scratch free, and punish it when it has an accident or goes to the wrong destination.
- Find a model that accurately predicts the stock market both on historical data, and on future data, and make lots of money with it.
- Beat anyone in the world at chess or the game of go.
- Analyze the human genome and find all genes complicit in cancer.

Find the perfect document on the Web that satisfies the user's intent as indicated by a query.

Clearly, these are all hard problems. They all share in the fact that one is searching for a solution that is hidden among an infinite number of inferior solutions. Not only is what we are looking for hidden, but it may also be hard to recognize it when put right in front of your face.

Search is such an interesting problem precisely because it resembles learning, reasoning, evolution, and other forms of deep and profound adaptation. The topics of neural networks, artificial intelligence, and genetic algorithms are all subsets of the general search problem.

We believe that the Web is rapidly approaching the point that it will be humanity's best effort of organizing the collective knowledge of all humanity. clearly surpasses the library of Alexandria and it will soon surpass the US Library of Congress and all other libraries in sheer size.

There is a mathematical result, published in 1997, due to David Wolpert and William Macready [Wolpert 1997], referred to as the 'No Free Lunch' theorem (or NFL for short) that is often misunderstood. The theorem deals with algorithms for solving the search problem. The theorem has both pessimistic and optimistic interpretations, hence the confusion surrounding it. The pessimistic interpretation can be summarized as:

All search algorithms are equally bad.

Put in this way, it should be clear why so many people are unhappy with it. In fact, if you are a scientist that has invested years showing that your type of search algorithm is better than most, than the NFL theorem is outright slanderous.

A more complete characterization of the NFL theorem would be:

Averaged over all possible spaces, even crazy ones that never occur in nature, all search algorithms are equally bad.

This alternative view clarifies the major caveat with the NFL theorem, namely, that it is making a statement about all search algorithms if they were applied to all possible search problems even ones that could never exist in our universe. There is still another way to characterize the NFL theorem, which we believe is both optimistic and realistic:

If your search algorithm is moulded to a particular problem space, it can work better than most other search algorithms.

The remaining caveat to this more gentle interpretation is that the penalty for being optimized to a particular problem domain is that the same solution that works well in one domain may prove horrible in every other domain. So it goes, we say.

All of this may seem to be completely unrelated to this chapter; however, we believe that the NFL theorem is key to understanding the current state of the Web and how it will evolve over time. In a nutshell, our claim is that the Web has co-evolved with humanity, and it will continue to do so. Moreover, we believe that the Web will approach a level of complexity that is on par with all human culture and with the human mind.

Simplicity, Complexity, and Search

Much of this chapter has focused on how the Web possesses an amazing array of properties that smack of both simplicity and complexity. To better appreciate this point consider how complicated a miniature version of the Web could be with only ten pages.

If each page is permitted to link to any of the ten, including itself, then there are 2^{100} different ways in which to connect up ten pages. This number is larger than the number of electrons in the universe. Now, instead of ten pages, think about billions and consider how complicated the Web could be if authors, pages, and users were not so regular in their collective behaviours. A billion pages with links pointing everywhere would truly be intractable and effectively unimaginable because no one would be able to make any sense out of it.

The point of this exercise is simple: the Web could have been tremendously complex, but it is not. In fact, the Web is exceedingly regular given its size and the lack of central authority. Moreover, this regularity can be exploited to make more effective algorithms for finding information on the Web.

Recall from the previous section our discussion of the PageRank algorithm. PageRank is mathematically very well understood. As an algorithm, it performs an iterative calculation that must be repeated multiple times, and the required number of iterations is easily known in terms of the error rate (associated with not running it for an infinite number of steps) and the properties of the link structure of the Web.

If the Web was not self-organized and if the link structure did not follow a power law, in all likelihood

PageRank would not be a practical algorithm because its required number of iterations would be close to infinite. Instead, we know that the Web is a forgiving domain for PageRank, in the sense that its power law properties all but guarantee that PageRank will quickly converge to valuable results.

This is an extremely subtle but important point: the Web's self-organized and fractal properties make it easier for algorithms to make sense of it. Moreover, these self-organized and fractal properties are a direct consequence of our (humanity's) own self-organization and fractal nature.

The Future

We believe that the Web is rapidly approaching the point that it will be humanity's best effort of organizing the collective knowledge of all humanity. It clearly surpasses the library of Alexandria and it will soon surpass the US library of Congress and all other libraries in sheer size.

The Web will continue to become an integral part of society, nearly blending into the background, as much of our society transitions into a dual nature that includes both a physical and a virtual existence.

We also believe that the generalized search problem – and the problem of building a nearly perfect search engine, in particular – will increase in importance as the need to find information on the Web becomes more ubiquitous and necessary to our day-to-day lives. In the future, Web search engines will radically change, ultimately possessing enough intelligence to simultaneously recognize the needs of the users that use it while making sense of the plethora of available information.

In short, we believe that the Web will become a mirror to humanity in the aggregate, and that the search engine will become a mirror to the human mind, and it is the self-organized and fractal nature of the Web that is both a symptom and a cause for this co-evolution.

Acknowledgements

We thank Bill Cheswick, KC Claffy, Roger Corn, William Decker, Dan Fain, Young Hyun, Steve Lawrence, Will Leland, Margaret Murray, and William Pryor.

References

Albert-Lászlo Barabási and Réka Albert. Emergence of scaling in random networks. *Science*, 286: 509–512, 1999.

Sergey Brin and Lawrence Page. The Anatomy of a Large-Scale Hypertextual Web Search Engine. *Proceedings of the 7th World Wide Web Conference (WWW7)*, 1998.

Andrei Broder, Ravi Kumar, Farzin Maghoul, Prabhakar Raghavan, Sridhar Rajagopalan, Raymie Stata, Andrew Tomkins, and Janet Wiener. Graph structure in the web: Experiments and models. *Proceedings of The 9th World Wide Web Conference (WWW9)*, 2000.

Soumen Chakrabarti, Mukul M. Joshi, Kunal Punera, and David M. Pennock. The structure of broad topics on the Web. *Proceedings of The 11th World Wide Web Conference*, 2002.

Mark E. Crovella and Azer Bestavros. Self-similarity in World Wide Web traffic: Evidence and possible causes. *Proceedings of ACM SIGMETRICS*, 1996.

Mark E. Crovella, Murad S. Taqqu, and Azer Bestavros. Heavy-tailed probability distributions in the World Wide Web. In *A Practical Guide To Heavy Tails*, chapter 1, pp. 3–26, Chapman & Hall, New York, 1998.

Martin Dodge and Rob Kitchin. *Atlas of Cyberspace*. Addison-Wesley, 2002.

Stephen Dill, Ravi Kumar, Kevin McCurley, Sridhar Rajagopalan, D. Sivakumar, and Andrew Tomkins. Self-similarity in the Web. *Proceedings of International Conference on Very Large Data Bases*, pages 69–78, 2001.

Michalis Faloutsos, Petros Faloutsos, and Christos Faloutsos. On power-law relationships of the Internet topology. *Proceedings of ACM SIGCOMM, 1999.*

Gary William Flake, Steve Lawrence, and C. Lee Giles. Efficient Identification of Web Communities. *Proceedings of ACM SIGKDD 2000.*

Gary William Flake, Steve Lawrence, C. Lee Giles, and Frans M. Coetzee. Self-organization of the Web and identification of communities. *IEEE Computer,* 35(3):66–71, 2002.

M. R. Gary and D. S. Johnson. *Computers and Intractability: A Guide to the Theory of NP-Completeness.* W. H. Freeman, New York, 1979.

Steven D. Gribble, Gurmeet Singh Manku, Drew Roselli, Eric A. Brewer, Timothy J. Gibson, and Ethan L. Miller. Self-Similarity in File-Systems, *Proceedings of ACM SIGMETRICS,* pp. 141–150, 1998.

Bernardo A. Huberman, Peter L. T. Pirolli, James E. Pitkow, and Rajan M. Lukose. Strong regularities in World Wide Web surfing. *Science,* 280(5360): 95–97, 1998.

Jon Kleinberg. Authoritative sources in a hyper-linked environment. *Journal of the ACM,* 46(5):604–632, 1999.

Jon M. Kleinberg, Ravi Kumar, Prabhakar Raghavan, Sridhar Rajagopalan, and Andrew S. Tomkins. The Web as a graph: Measurements, models, and methods. *Proceedings of the 5th International Conference on Computing and Combinatorics,* pp. 1–18, 1999.

Jon Kleinberg and Steve Lawrence. The structure of the Web. *Science,* 294: 1849–1850, 2001.

Ravi Kumar, Prabhakar Raghavan, Sridhar Rajagopalan, and Andrew Tomkins. Trawling the web for emerging cyber-communities. *Proceedings of the 8th World Wide Web Conference (WWW8),* 1999.

Will E. Leland, Murad S. Taqqu, Walter Willinger, and Daniel V. Wilson. On the self-similar nature of Internet traffic. *Proceedings of ACM SIGComm,* 1993.

Aristid Lindenmayer, Mathematical models for cellular interaction in development, parts I and II, *Journal of Theoretical Biology,* 18, 1968.

Benoît Mandelbrot. An informational theory of the statistical structure of language. In *Communications Theory* (Willis Jackson, ed.), Academic Press, New York, 1953.

Benoît Mandelbrot. A note on a class of skew distribution functions: Analysis and critique of a paper by H. A. Simon. *Information and Control,* 2:90–99, 1959.

R. M. May. How many species are there on earth? *Science,* 214: 1441–1449, 1988.

Daniel Menascé, Bruno Abrahão, Daniel Barbará, Virgílio Almeida, and Flávia Ribeiro. Fractal characterizations of Web workloads. *Proceedings of the 11th World Wide Web Conference, Web-Engineering Track,* 2002

David M. Pennock, Gary William Flake, Steve Lawrence, Eric J. Glover, C. Lee Giles, Winners don't take all: Characterizing the competition for links on the web. *Proceedings of the National Academy of Sciences,* 99(8):5207–5211, 2002.

James E. Pitkow. Summary of WWW characterizations. *Computer Networks and ISDN Systems,* 30:(1–7):551–558, 1998.

Manfred Schroeder. *Fractals, Chaos, Power Laws: Minutes from an Infinite Paradise,* W H Freeman & Co., 1995.

Herbert A. Simon. On a class of skew distribution functions. *Biometrika,* 42: 425–440, 1955.

Duncan J. Watts and Steven H. Strogatz. Collective dynamics of 'small world' networks. *Nature,* 393: 440–442, 1998.

David H. Wolpert and William G. Macready. No free lunch theorems for search. *IEEE Transactions on Evolutionary Computation,* 1, 1997.

George Kingsley Zipf. *Human Behavior and the Principle of Least Effort.* Addison-Wesley Press, Massachusetts, 1949.

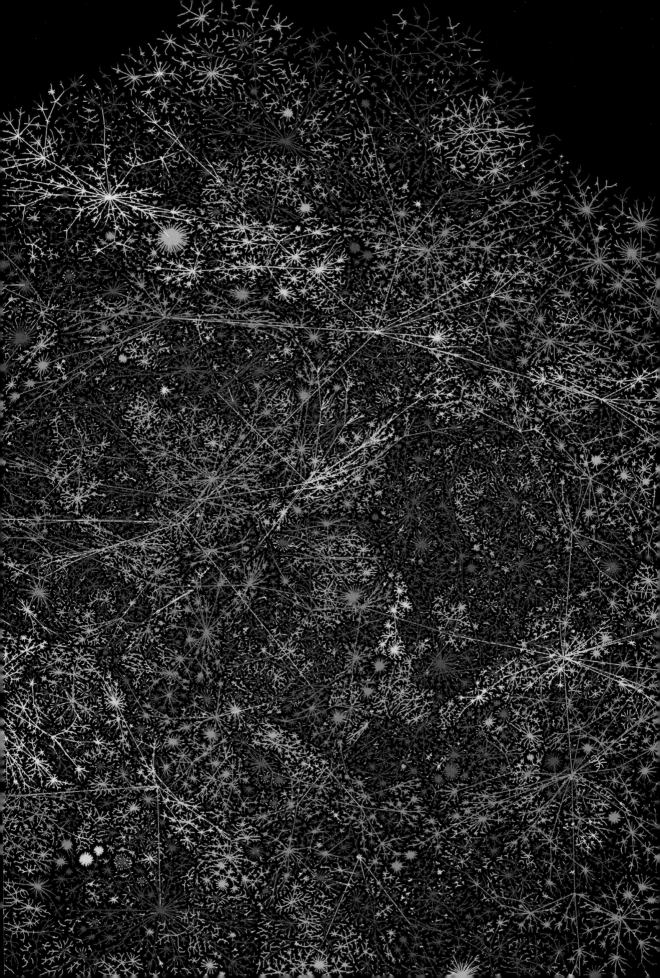

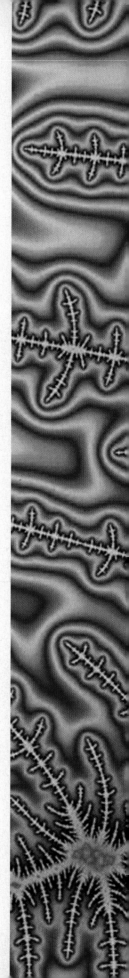

7 Fractal Financial Fluctuations

Benoît Mandelbrot

In the great depression of the 1930s the economist John Maynard Keynes wrote: *'We have involved ourselves in a colossal muddle, having blundered in the control of a delicate machine, the workings of which we do not understand.'* In this century the very same machine is far bigger and much more delicate and we continue to struggle desperately to understand just how it works.

The collapse of our banks, hedge funds and other lenders – as well as rising unemployment – leave us gasping for air, comprehension and reassurance. As we flounder it would be foolhardy not to investigate models providing more accurate estimates of risk. We do thankfully find that fractal and multifractal geometry throws considerable light on our ever-darkening puzzle.

A discipline better known as describing the shapes of coastlines and clouds and the distribution of galaxies and as having led to the discovery of the Mandelbrot set, this geometry has also been successful in describing the growth and collapse of financial prices. (Nigel Lesmoir-Gordon)

Benoit Mandelbrot is unusual in having left a permanent mark on numerous and seemingly diverse fields of science and art. He is Sterling Professor Emeritus of Mathematical Sciences at Yale University. In 1958 he had joined IBM's Thomas J. Watson Research Center, where he served as IBM Fellow in Physics, 1974–1993. He belongs to the National Academy of Sciences of the USA, the American Academy of Arts and Sciences, the Norwegian Academy of Science and Letters, and the American Philosophical Society. Having attended Ecole Polytechnique and earned the Ph.D. from the University of Paris, he also holds numerous honorary doctorates from institutions throughout the world. He received the 1993 Wolf Prize for Physics and also (among others) the Barnard, Franklin and Steinmetz Medals and the Harvey, Humboldt, Nevada, Honda and Art and Science Prizes.

N. Lesmoir-Gordon (ed.), *The Colours of Infinity: The Beauty and Power of Fractals*, DOI 10.1007/978-1-84996-486-9_7, © Springer-Verlag London Limited 2010

Were all the fountains of the great deep broken up, and the windows of heaven were opened. And the rain was upon the earth forty days and forty nights. (Genesis: 6, 11–12)

There came seven years of great plenty throughout the land of Egypt. And there shall arise after them seven years of famine. (Genesis: 41, 29–30)

These two quotes from the Bible (In King James translation) suffice to show that Man has known for millennia that the Earth's physical environment is subject to ferocious variations of at least two very separate kinds. The intensity of those fluctuations has, however, long been underestimated in the Sciences, hence not faced properly.

The same is true of the historical fluctuations on competitive markets, which are not ruled by physics, rather by the mostly unknown laws of financial economics. As is being written, early in 2009, screaming headlines bear witness that the unevenness, the "instability," of both Nature and Culture has been very broadly and deeply underestimated.

Seeking a closer relation between theory and fact, I have been studying such "wild" fluctuations for about forty years. My extensive work on price variation may have been overshadowed – for example – by the Mandelbrot Set, but current events (early in 2009) demonstrate its significance. The present paper is an informal introduction.

Motivated by the two quotes from the Bible with which this piece began, my terminology classifies the threatening deviations in two categories. Because of the Biblical story of the Flood and Noah, "Noah Effect" will denote major changes that occur rapidly (even instantaneously) but have strong and durable consequences. Because of the Biblical story of the dream of Pharaoh and Joseph the son of Jacob, "Joseph Effect" will denote sequences of changes that need not seem threatening when viewed individually but have major cumulative effect.

The Noah and the Joseph obstacles to sustainability range from the wholly natural (earthquakes, volcanoes, non man-made climate changes) to the wholly man-made. Both "effects" have long characterized the phenomena I have chosen to study most. I shall make minimal reference to the other aspects of the Noah and Joseph Effects, and focus on one field of application.

That field is the variation of financial prices, that is, of prices quoted on financial markets that trade securities, commodities, and exchange or interest rates. The wild volatility of those markets has long been known. The details belong to political economy and will be briefly touched upon at the end of this chapter. Here, the aim of this chapter is to further the knowledge of the underlying facts and to contribute to a better understanding of price variation. Inevitably, it criticizes previous strongly held views on this topic, particularly, the "coin tossing" or Brownian model. Unfortunately and perhaps surprisingly, the existing models of financial price variability are not good enough and new research is keenly needed, because the points that matter most have, in previous research, been deliberately set aside or disregarded.

The point of departure is that financial prices, including those of securities, commodities, foreign exchange or interest rates, are largely unpredictable. The best one can do is to evaluate the odds for or against some desired or feared outcomes, the most extreme being "ruin." Those odds will also be used as inputs for decisions concerning economic policy or changes in institutional arrangements. To handle all those issues, the first step – but far from the last! – is to represent different instances of price variation by suitable random processes.

The word "suitable" and the plural in "processes" will surprise many readers. It is, indeed, widely believed that "random change" is a synonym for "prices that move up a bit or down a bit following the toss of a coin." The technical term is "simple random walk." It was made popular by a book title that won the high distinction of becoming a cliché, namely, "random walk down the Street."

The belief that there is no alternative is strengthened by the fact that – in effect – the coin tossing model dates to 1900! It is, indeed, by far the oldest and the most widely used model of price variation. The (unsaid) point of departure of this chapter is that the term, random, has a far broader meaning, allowing the coin tossing model to be replaced by alternatives. It will be argued that the "multifractal" alternative that I put forward is very suitable indeed.

The multifractal model does not belong to esoteric mathematics and it must not be allowed to remain part of pure science. Its practical consequences are many and very serious. The first but not last is in the spirit of the Hippocratic Oath, "do no harm," which deserves to be generalized to finance and is best expressed in nautical terms. When a ship was built to navigate placid lakes by fair weather, to send it across the ocean in typhoon season is to do serious harm. Similarly, the "coin tossing" model of financial prices (and its kin) may well be beloved by mathematicians, but it denies the existence of hurricanes; therefore it is dangerous.

The preceding nautical analogy will be heavily used throughout this text, because it resides at the very centre of the present study. Many alternatives to the coin tossing model are available, but the multifractal alternative differs from the others in "qualitative" ways that have immediate consequences for finance and economic policies.

The coin tossing model exemplifies a form of randomness (a "state of randomness," as I shall argue) that can be called "mild." Had the evidence agreed with this model – but it does not at all – variability in finance would be as easily controllable as is variability in physics.

However, the coin tossing model must not be criticized too hard. It is always best to start with the simplest possible model and hold to it until it has begun to bring more harm than value. In its time, it played a fundamental and positive role in creating awareness of the difficulty of even the simplest forms of randomness. One can also argue that for the "man in the street" coin tossing is an adequate description of the facts. But policy makers and the professionals in finance are (or should be) far more demanding for them, as will be seen. It matters very much that coin tossing is very far from accounting for some essential facts.

Once again, the history of price variation is filled with "financial hurricanes" while we shall see that coin tossing claims that they practically never happen. Shipbuilders and ship owners cannot predict the dates and destructiveness of the hurricanes their vessel will encounter over its lifetime. But the knowledge that hurricanes will happen – and realistic evaluation of the corresponding odds – permeates ship-building, ship ownership and navigation.

This chapter argues that tools needed to acquire some mastery of the intensities of financial hurricanes are already available. They are those of fractal and multifractal geometry, a discipline better known as describing the shapes of coastlines and clouds and the distribution of galaxies and as having led to the discovery of the Mandelbrot set. My claim is that it also describes the growth and collapse of financial prices.

Pick the Fakes

This chapter includes a multitude of words or numbers and also of formulas and dry diagrams. Without mastering them, my claims and contributions cannot be fully understood and appreciated. But to make the central point, words and formulas are not really necessary if one uses diagrams.

To explain this, the best and quickest way is to encourage the reader to participate in a test concerning Figs. 7.1 and 7.2. No one is asked to accept pictures as the sole or final arbiter in scientific discourse, only as a useful additional tool. Pictures are often used to delude, but in this instance they deserve to be described as uncovering a widespread delusion and assisting in the selection of an improvement.

Drawn in no particular order, some of the graphs in Fig. 7.1 are "real" plots of the behaviour in time of some actual financial prices. Other graphs are computer drawn "forgeries" of the outputs of diverse models.

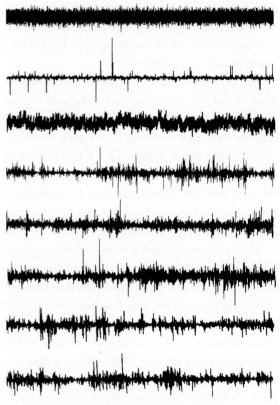

Fig. 7.2 A stack of diagrams, illustrating the successive "daily" differences in some actual financial prices and some mathematical models. The reader is challenged to pick the fakes.

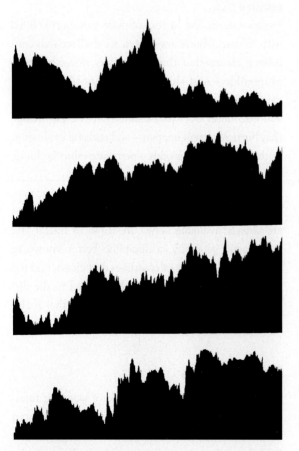

Fig. 7.1 A collection of diagrams, illustrating – in no particular order – the behaviour in time of some actual financial prices and of some mathematical models of this behaviour. It would be very difficult to pick the fakes.

The real ones follow the practice of financial journals and trace the sequence of daily closing of some price series such as security, commodity, foreign exchange, or interest rate. The "forgeries" correspond to more or less effective imitations of financial reality – provided by mathematical models that are fully specified in quantitative fashion, therefore can be sampled and illustrated without resorting to unreported stretching and reducing or other such manipulations. For each graph of Fig. 7.1, Fig. 7.2 plots the "price" differences from one day to next.

Now the "find-the-fakes test" can be described: you are asked to separate reality and forgery as completely as you can. For a perfect score, you must rank the diagrams from "most obviously a forgery" to "apparently real."

When the test relates to Fig. 7.1, all records look very much alike. To separate the real and forged records is very difficult. This impression is confirmed by looking through the financial press and the books on the mathematics of finance. The optimist will rush to conclude that coin tossing – which is represented by one of the graphs in Fig. 7.1 – is perfectly acceptable.

Unfortunately, as we shall see momentarily, this optimism would be seriously misplaced. The resemblance between those curves is due to the fact that, on graphs of prices themselves, important differences are not revealed or enhanced. They are hidden. In other words, plots of prices are a very misleading way of presenting information. This is well-known to students of the psychology of vision: position is seen less accurately than change.

In sharp contrast, the lines in the stack of this Fig. 7.2 strikingly differ from one another. The meaning of those differences will be refined through this chapter, which should end by revealing the solution of the test.

Large Stock Market Movements and Their Odds

Individual investors and professional stock and currency traders know better than ever that prices quoted in any financial market often change with heart-stopping swiftness. Fortunes are made and lost in sudden bursts of activity when the market seems to speed up and the volatility soars.

In September 1998, for instance, the stock for Alcatel, a French telecommunications equipment manufacturer, dropped about 40% one day and another 6% over the next few days. In a reversal, the stock shot up 10% on the fourth day. On a longer time scale most real price changes behave like those in the lower portion

of Fig. 7.2. However, not all lines at the bottom of Fig. 7.2 are real. (However, I am not giving away the test the reader is in the process of taking!)

The coin tossing model, which served as foundation for the theory of finance used most widely during the twentieth century, is represented by the top line of Fig. 7.2 (now, I am giving away part of the test). As we see, precipitous events like the Alcatel debacle are given totally negligible odds. Certainly, they should never happen in the lifetime of this generation and the next few. A cornerstone of finance is "modern portfolio theory," which tries to maximize returns for a given level of risk. The mathematics underlying portfolio theory ignore the possibility of a typhoon.

This term, coin-tossing, is actually an oversimplification However, the risk-reducing formulas behind portfolio theory rely on a number of demanding premises that are mathematically attractive but rely on hope rather than reality. First, they suggest that price changes are statistically independent of one another: for example, that today's price has no influence on the changes between the current price and tomorrow's. This is the "efficient market" hypothesis – attractive beyond words, but simply lethal.

The second assumption is that all price changes are distributed in a pattern that conforms to the standard "bell curve" of statistics. Of the three curves in Fig. 7.3,

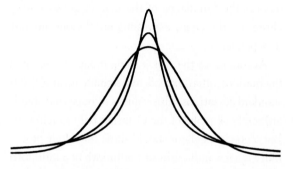

Fig. 7.3 Shapes of the Gaussian distribution and of two "stable" distributions. The latter provide a far better fit for many financial data, but the multifractal model is even more satisfactory.

the bell curve is the flattest in the centre. The width of its bell (as measured by its "sigma," or standard deviation) depicts how far price changes diverge from the mean. In this perspective, 95% of all cases fall into the narrow range between minus two sigmas and plus two sigmas. As was already mentioned and will be elaborated momentarily, the bell curve declares extreme events to be extremely rare. Typhoons are, in effect, defined out of existence.

Do financial data neatly conform to such assumptions? Of course, they *never* do! This is shown by a more attentive inspection of the bottom portion of Fig. 7.2. It is true that charts of stock or currency changes over time reveal a constant background of small up and down price movements – though not as uniform as one would expect of price changes that fit the bell curve. Invariably, however, these patterns constitute only one aspect of the graph. A substantial number of sudden large changes – spikes on the chart that shoot up and down as with the Alcatel stock – stand out from the background of more moderate perturbations.

But the presence of "long tails" is very far from being all. Equally important is the fact that successive price movements are not independent. It is typical of their magnitude – large or small – to remain roughly constant for an extended period and then suddenly and unpredictably increase **or** decrease for another extended period. Big price jumps become more common as the turbulence of the market grows – they cluster on the chart, expressing an obvious amount of dependence.

According to the coin tossing model, these large fluctuations often exceed ten sigmas, meaning ten standard deviations. This value is so huge that standard textbook tables of the Gaussian fail to include it. But a good calculator should show that their probability is a few millionths of a millionth of a millionth of a millionth, that is, approximately one day out of ten million million million years. If risks were so

tiny, they would not deserve even a passing thought. But this tiny value grossly contradicts the evidence. The real world of finance produces "ten sigma" spikes on a regular basis – as often as every month, to give an idea – and their probability should be expected to be a few hundredths.

The tiny probability mentioned a few lines above illustrates that the Gaussian practically vanishes near the left and right ends of the graph. Had the horizontal axis been part of Fig. 7.3 (which – by design – it is not), it would have hidden those insignificant tails.

Reality is incomparably better represented by the other two curves in Fig. 7.3, both with more peaked heads and fatter tails. These curves belong to the "M 1963 model" produced by my first attack on financial data. Having revealed this fact, it is best to narrow down the test the reader is supposed to be taking. Price changes according to the M 1963 model are the source of the second line in Fig. 7.2. This is clearly better than the top line, to be sure, but far from being the last word.

Coin Tossing Normality Versus the Financial Reality

The bell curve is often described as illustrating the "normal" distribution. But does it follow that financial markets should be described as "abnormal or anomalous"? Of course, not. They are what they are, and it is the coin tossing model, and therefore portfolio theory, which is flawed.

Modern portfolio theory poses a danger to those who believe in it too strongly and is a powerful challenge for the theoretician. The extremely bearish answer acknowledges faults in the present body of thinking, yet claims that there is no alternative: that very large market swings are anomalies, individual "acts of God" that present no conceivable regularity. Other adherents suggest that no other premise can both be handled through mathematical modelling and lead to a rigorous quantitative description of at least some features of major financial upheavals. In the absence of an alternative, coin tossing must be maintained, "faute de mieux."

An increasingly wide agreement is being reached that this extremely bearish view is untenably bad science and that the coin tossing model must be replaced by one that allows (near-)instantaneous price changes and substantial temporal dependence. This agreement marks a change of mood in the "mainstream," bringing it toward the views I have been campaigning for since 1963 and 1965, respectively. From this point on, however, two general approaches are in conflict, leading to what I shall call "micromanaged" and "macromanaged" models.

Micromanaged models agree with my diagnosis but not my follow-up. They proceed through a series of "fixes." Each fix "patches" a perceived defect of coin tossing, independently of its other defects. The outcome is that this approach accumulates a large number of parameters and no property is present that was not knowingly incorporated in the construction. In the nautical analogue, the fixes consist in lengthening a small boat's keel, lengthening its mast, reinforcing its engine, etc..., one by one. My experience of successful modeling in other fields has fostered deep a priori doubts about the chances of micromanaged modeling in finance. But personal prejudices would not have mattered if a posteriori modeling had been effective. I think it has not.

My own work – carried out over many years – takes the very different and decidedly bullish position that it is clearly preferable to design a large boat from scratch. I claim that a financial model can be redesigned following an approach that is macromanaged by being guided by a principle of fractal invariance, to which we shall come soon. The outcome, as I propose to show, is that the variation of financial prices can be accounted for by a model derived from my work in fractal geometry in an elaboration called multifractals. Once again, I never claimed the ability to predict the future with certainty. But multifractals do create a more realistic picture of market risks. Given the recent events (this is written in January, 2009) it would be foolhardy not to investigate models providing more accurate estimates of risk.

Fractals, Multifractals and the Market

An extensive mathematical basis already exists for fractals and multifractals. Fractal patterns do not only appear in the price changes of securities but also in the distribution of galaxies throughout the cosmos, in the shape of coastlines and in the decorative designs generated by innumerable computer programs.

A fractal is a geometric shape with the property that it can be separated into parts, each of which is a reduced-scale version of the whole and has that same property itself. (The trouble is that by this definition an interval or square are fractals. This problem is familiar in a classical context, insofar as the definition of complex numbers does not prevent them from being real.) In finance, this concept is not a rootless abstraction but a theoretical reformulation of a down-to-earth bit of market folklore, namely, the notion that movements of a stock or currency all look alike when a market chart is enlarged or reduced so that it fits some prescribed time and price scales. This implies that an observer cannot tell which of the data concern price changes from week to week, day to day, or hour to hour. This quality defines the charts as fractal curves and many powerful tools of mathematical and computer analysis become available.

A technical term for this form of close resemblance between the parts and the whole is self-affinity. This property is related to the better-known concept of fractals called self-similarity, in which every feature of a picture is reduced or blown up by the same ratio, a process familiar to anyone who ordered a photographic enlargement or a photocopy. Financial market charts, however, are far from being self-similar. If we simply focus on a detail of a graph, the features become increasingly higher than they are wide – as are the individual up-and-down price ticks of a stock. Hence, the transformation from the whole to the parts must shrink the time scale (the horizontal axis) less than the price scale (the vertical axis). This task is routinely performed by copiers using lasers. The geometric relation of the whole to its parts is said to be one of self-affinity.

Unchanging properties are not given much weight by most statisticians, but they are beloved of physicists and mathematicians like myself. We call them invariances and are happiest with models that present an attractive invariance property. A good idea of what I mean is provided by drawing a simple chart that inserts (interpolates) price changes from time 0 to a later time 1 in successive steps. The intervals themselves are chosen arbitrarily; they may represent a second, an hour, a day or a year.

The process begins with a price represented by a straight trend line called "initiator," shown in the top panel of Fig. 7.4. Next, a broken line called "generator" is used to create the pattern that corresponds to a slow up-and-down price oscillation. It is obviously essential that the number and positions of the pieces of the generator are completely specified. As soon as one allows oneself the right to fiddle with the generator during the construction, no prediction can be made.

In Fig. 7.4, the generator consists of three pieces that are inserted (interpolated) to refine the straight trend line. A generator with fewer than three pieces could not simulate a price that must be able to move up and down. Then each of the generator's three pieces is interpolated by three shorter ones.

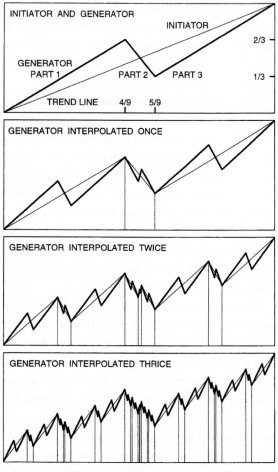

Fig. 7.4 Constructing a "pseudo-Brownian cartoon" of the idealized coin-tossing model that underlies modern portfolio theory. The construction starts with a linear trend ("the initiator") and breaks it repeatedly by following a prescribed "generator." The interpolated generator is inverted for each descending piece. The pattern that emerges increasingly resembles market price oscillations.

Repeating these steps reproduces the shape of the generator, or price curve, but at increasingly compressed scales. Both the horizontal axis (time scale) and the vertical axis (price scale) are squeezed to fit the horizontal and vertical boundaries of each piece of the generator.

Interpolations Continued (Not Quite) Forever

Only four construction stages are shown in Fig. 7.4, but the same process continues. In theory, it has no end, but in practice, it makes no sense to interpolate down to time intervals shorter than those between transactions, which may be of the order of a minute. The fact that each piece ends up with a shape like the whole is not a surprise: this scale invariance is present simply because it was built in. The novelty (and surprise) is that these very simply defined self-affine fractal curves suffices to exhibit a wealth of structure. The beauty of fractal geometry is that it does not consist in micromanaged models in which everything of interest has been inputed separately. Fractals involve only macromanaged instructions, yet yield models general enough to reproduce the patterns that characterize portfolio theory's placid markets as well as the tumultuous trading conditions of real markets. Indeed the construction's outcome, if plotted as in Fig. 7.2, is very sensitive to the exact shape of the generator.

For example, Fig. 7.4 uses a very special generator that – according to a theory I developed – will produce a behaviour that is pseudo-Brownian, meaning that it is close to the relatively tranquil "mildly random" picture of the market ruled by coin tossing. But this level of tranquillity prevails only under extraordinarily special conditions that are satisfied only by equally special generators. Figure 7.3 satisfies those conditions because each generator segment's height – namely, 2/3, 1/3 or 2/3 – was made equal to the square root of the corresponding width – namely, 4/9, 1/9 or 4/9. This "square root rule" is a characteristic of a process physicists call "simple diffusion." Adherence to the assumptions behind this oversimplified model is one of the central mistakes of modern portfolio theory. It is much like a theory of sea waves that forbids the swells to exceed six feet.

A first and very important generalization of Fig. 7.4 yields models that are non-Brownian but can be called "unifractal." It consists in continuing to require that the height of every segment of the generator be linked to its width by the same relation in the form of a power H. Previously, we set H = 1/2 , but a different value of H can be chosen, as long as it is positive and less than 1. Taking H = .7 suffices to change the top line of Fig. 7.2 into its third line.

On the corresponding graphs in Fig. 7.1, the place of tranquility and mildness is taken by movements that are non-periodic but described by everyone as "cyclic" with many apparent cycle lengths, ranging from very small up to "about three cycles in a sample." (This last rule is a remarkable observation that cannot be elaborated here.) Here, cyclic behaviour is present in the output without having been incorporated in the input. This is lovely, but large spikes of variation were lost and must be reinstated.

There is a second and far more drastic generalization of Fig. 7.3. So far, market activity was assumed constant but one can allow it to speed up and slow down. This variability is the essence of volatility. In fact, practical people describe the diverse lines at the bottom of Fig. 7.2 as proceeding at many different speeds and at different times. This is why models that allow for variability add the prefix "multi" before the word "fractal."

To define "activity" is beyond our concern and not necessary. The key idea is that the market does not follow the physical time that proceeds with the relentless regularity of a clock, but instead a subjective time that flows slowly during some periods and fast during others.

In this spirit, the theory provides a neat "transmutation" from uni to multifractal. The key step shown in Fig. 7.5 is to lengthen or shorten the horizontal time axis so that the pieces of the generator are either stretched or squeezed. At the same time,

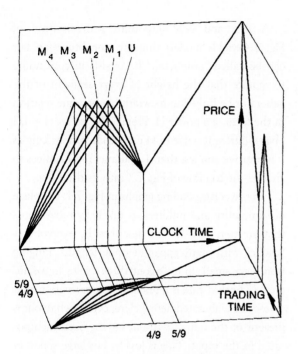

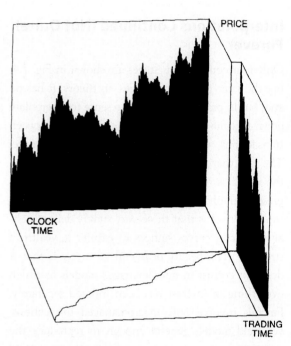

Fig. 7.5 This open cube illustrates related generators: The "right wall" shows an oscillating generator in trading time. This is the pseudo-Brownian (unifractal) generator of Fig. 7.4. The "back wall" shows four multifractal oscillating generators in clock time. The "floor" shows the generators that relate the clock time to trading time. Each is an increasing function of the other. Moving a piece of the fractal generator to the left causes the same amount of market activity in a shorter time interval for the first piece of the generator and the same amount in a longer interval for the second piece.

Fig. 7.6 The underlying pattern is as in Fig. 7.5, but limited to the left-most generator, and the generators are replaced by the curve obtained by repeating it recursively as done in Fig. 7.3. To make the picture clearer, the back and right wall are moved away from the floor.

the vertical price axis may remain untouched. As seen on the "back" wall of Fig. 7.5, the first piece of the unifractal generator is progressively shortened, which also provides room to lengthen the second piece. After making these adjustments, the generators become multifractal (M1 to M4). As seen on the "floor," of Fig. 7.6, market activity speeds up in the interval of time represented by the first piece of the generator and slows in the interval that corresponds to the second piece.

When the generators in Fig. 7.5 are used recursively, one obtains the patterns shown in Fig. 7.6.

Recall that those patterns do not pretend to exhaust all the possibilities offered by either theory or the facts. Their sole aim is to show the power of the very simplest fractal models.

Such an alteration to the generator can produce a full simulation of price fluctuations over a given period, using the process of interpolation described earlier. Each time the first piece of the generator is further shortened. The process of successive interpolation produces a chart that increasingly resembles the characteristics of volatile markets (Fig. 7.7).

Once again, the unifractal (U) chart that prevails before any shortening corresponds to the becalmed markets postulated in the portfolio theorists' model. Proceeding down the stack (Ml to M4), each chart

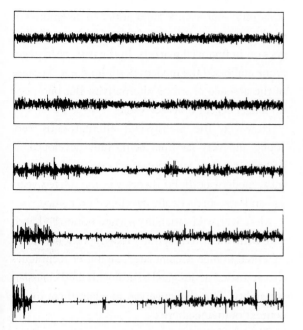

Fig. 7.7 Randomized multifractal "price increments" that correspond to the five multifractal generators in Fig. 7.5. On top a pseudo-Brownian sequence of "price increments". A gradual displacement of the generator to the left causes market activity to increase gradually, becoming more and more volatile.

diverges further from that model, exhibiting the sharp, spiky price jumps and the persistently large movements that characterise financial trading.

To make these models of volatile markets achieve the necessary realism, the figures involve an important detail, which has not been mentioned yet. The three pieces of each generator were scrambled – a process not shown in the illustrations. It works as follows. Altogether, the three pieces of the generator allow the following six permutations:

1,2,3; 1,3,2; 2,1,3; 2,3,1; 3,1,2 and 3,2,1.

Conveniently enough, a die has six sides; imagine that each bears the image of one of the six permutations. Before each interpolation, the die is thrown and the permutation that comes up is selected.

Back to the Game of Picking the Fakes

How do simulations of the multifractal model stand up against actual records of changes in financial prices? To respond, let us return to Fig. 7.2, which is a composite of several historical series of price changes with a few outputs of artificial models.

As we have already observed the goal of modelling the patterns of real markets is certainly not fulfilled by the first chart, which is extremely monotonous and reduces to a static background of small price changes, analogous to the static noise from a radio. Volatility stays uniform with no sudden jumps. In a historical record of this kind, daily chapters would vary from one another, but all the monthly chapters would read very much alike.

The rather simple second chart is less unrealistic, because it shows many spikes; however, these are isolated against an unchanging background in which the overall variability of prices remains constant. The third chart has interchanged strengths and failings, because it lacks any precipitous jumps.

The eye tells us that these three diagrams are unrealistically simple. Let us now recall the sources. Chart 1 illustrates price fluctuations in a model introduced in 1900 by French mathematician Louis Bachelier. The changes in prices follow a "random walk" that conforms to the bell curve and illustrates the model that underlies modern portfolio theory. Charts 2 and 3 are partial improvements on Bachelier's work: one is the "M 1963" model, which I proposed in 1963 (based on Levy stable random processes). The other is the "M 1965" model, which I published in 1965 (based on fractional Brownian motion). These revisions of coin tossing are inadequate, except under certain special market conditions.

By now, the test around which this chapter is structured has been reduced to a careful inspection of the more important five lower diagrams of the graph.

Let me now add a piece of information that was withheld until now: at least one record is real and at least one is a computer-generated sample of my latest multifractal model. The reader is free to sort those five lines into the appropriate categories. I hope the forgeries will be perceived as surprisingly effective. In fact, only two are real graphs of market activity. Chart 5 refers to the changes in price of IBM stock and chart 6 shows price fluctuations for the dollar/deutschemark exchange rate. The remaining charts (4, 7 and 8) bear a strong resemblance to their two real-world predecessors. But they are completely artificial.

Two technical points must be mentioned before moving on to conclusions. The recursive constructions in the body of the chapter were nothing but "cartoons". The artificial charts 4, 7 and 8 were, instead, generated through a refined form of my multifractal model, called "fractional Brownian motion in multifractal trading time." Secondly, this introductory survey necessarily emphasizes graphics, but – once again – the theory of multifractals is endowed with full numerical tools of analysis.

Very Tentative Conclusions: Diversification and Reinsurance

What conclusions should be drawn from all this? Does this matter to a corporate treasurer, currency trader or other market strategists? Does this matter to the central banker and others concerned with overall financial and economic policy? Does this matter to the economist who seeks to explain the workings of the economy and concedes that his task may be helped by an accurate description of part of what is to be explained?

All those questions arise because the discrepancies between coin-tossing and the actual movement of prices have become too obvious to be ignored much longer. Prices do not vary continuously, and they are subjected to wild fluctuations at all time scales.

Volatility – far from a static entity to be ignored or easily compensated for – is at the very heart of what goes on in financial markets. In the past, nearly everyone embraced the modern portfolio theory because of the absence of strong alternatives. But one need no longer accept it at face value.

However, the multifractal alternative is very young and very far from being fully developed. It deserves to draw attention (and criticism). By contrast, modern portfolio theory was formulated years ago and was energetically developed ever since.

Moreover, wild variability is a new notion endowed with little inherited capital. Modern portfolio theory inherited a large accumulated capital of techniques that statisticians designed to deal with mild Gaussian variability. The challenge was to adapt them to the context of financial prices.

Therefore, it is necessary, as we near a conclusion, to separate thoughts concerning the near future from thoughts concerning the longer range. Multifractals can immediately be put to work to "stress-test" portfolios, in particular, from the viewpoint of a quantity called "value at risk," whose definition is unfortunately beyond the scope of this chapter. Stress-testing begins by questioning how a portfolio would have performed if it had been designed a while ago. That is, the simplest stress test merely uses historical data. But the actual market test will not come in the past, but rather in the future, and a future that simply repeats the past is only one of many alternatives, and not a very likely one.

The goal of every model of price variation (coin-tossing not being an exception) is to use the past to create the same patterns of variability as do the unknown rules that govern actual markets. This attempt should yield a collection of alternative scenarios for the future, and stress-testing should include tests based on many such alternatives.

According to the coin-tossing model, the differences between those alternatives are comparatively slight. Not so with the multifractal models. They describe the past market fluctuations more realistically and the scenarios they propose for the future include a quota of extreme events that will really stress a portfolio. This is all that can be said on this subject at this point.

Of greatest interest, at least to me, are problems that the multifractal model confronts on broader institutional, temporal and spatial horizons. They are more important than any detail and question the worth of the widespread faith in the power of diversification and other forms of lumping and averaging. Here an enlightening analogy and powerful guidance for the future is provided by a distinction between different levels of insurance that relates to my distinction between mild and wild "states" of chance.

Most life, automobile, or homeowner risks are mild. Very much like the coin-tossing model of price change, they fall within a narrow range and are mutually independent. Even when a portfolio of insurance contracts is small, the wonders of diversification (due to the law of large numbers and related mathematics) can be trusted to create a risk of ruin that is sufficiently small to be profitable even for an insurance company with limited reserves. To play safe and to insure the occasional higher risk, the insurer of mild risks will seek reinsurance – which will seldom be needed, therefore will not be expensive. When a tornado defeats diversification of homeowner policies the reinsurer is likely to be an entity that had collected no premiums, namely an agency of a state.

However, many other risks seeking to be insured are wild, very much like in my multifractal model of price change. They involve the equivalents of the notorious "ten sigma" price changes that were discussed earlier in this chapter. Ordinary diversification would be defeated by such risks, even if the number of cases had sufficed for a law of large numbers to apply. More precisely, the odds of those wild risks, if included in the usual calculations,

would imply reserves that are clearly unreasonable. However, such risks become insurable if they are immediately shared with reinsurers (or almost directly with competitors, as is apparently the case in the shipping industry).

The key fact is that insurers cannot survive by only considering the "fair weather" 95% of the claims, which would have easily been diversified. Not only can the 5% of large "foul weather" claims not be ignored, but their odds are non-negligible and are an essential input of planned and carefully priced reinsurance.

Once again, theories based on coin-tossing legislate this "foul weather" out of existence, but it is evident that many features of the real world are best understood as designed to tackle comparatively rare but potentially disastrous situations. It is indeed filled with state or private institutions and informal or ad-hoc arrangements, whose purpose can be viewed as that of reinsurance. A central part of my thinking in finance is that those arrangements may have worked in the past but cannot be relied upon in the future. As for institutions, their role deserves a fresh examination.

As a result of globalisation and of the events of 2006, the relevance of the preceding comments on insurance is bound to increase. Under the coin-tossing model, the effects of globalisation are limited. But the actual behaviour of financial prices confirms what intuition suggests: the larger the markets, the greater the attention demanded by the potentially disastrous effects of financial storms.

To conclude, no overall mathematical technique comes close to forecasting a price drop or rise on a specific day on the basis of past records. Multifractals do not claim to do any better. But they provide estimates of the probability of what the market might do in the future and allow one to prepare for inevitable sea changes. The new modelling techniques are designed to cast a light of order into the seemingly impenetrable thicket of the financial markets. They also recognize the mariner's warning that, as recent events demonstrate, deserves to be heeded: On even the calmest sea, a gale may be just over the horizon.

Brief Bibliography Concerning Finance

B. Mandelbrot, Fractals and Scaling in Finance: Discontinuity, Concentration, Risk, Springer-Verlag, 1997.

B. Mandelbrot, Multifractals and 1/f Noise: Wild Self-Affinity in Physics, Springer-Verlag, 1999. Translations in numerous languages.

B. Mandelbrot, L. Calvet & A. Fisher, The Multifractal Model of Asset Returns, Three reports of the Cowles Foundation for Economic Research of Yale University.

B. Mandelbrot & R. L. Hudson, The (Mis)Behaviour of Markets: A Fractal View of Risk, Ruin, and Reward, Basic Books, 2009. Translations in numerous languages.

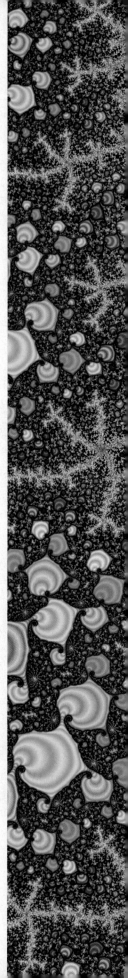

8 Filming The Colours of Infinity

Nigel Lesmoir-Gordon

Film-maker Lesmoir-Gordon offers a fascinating behind-the-scenes account of how a modern cult classic came into being. The task of moving from a 'Eureka!' moment of understanding the Mandelbrot Sets for the first time to making a compelling and saleable film about them was beset by challenges, on technical, logistical and financial levels, but was also helped by many strokes of good fortune. The sustaining vision, in the director's own words, was to 'set out to make, not a deeply mathematical, analytical piece, but rather a celebration of a remarkable discovery. I wanted it to be fun. I knew it had to entertain at some level if it was going to reach a really wide audience.' This chapter is followed by a transcript of the final script of The Colours of Infinity.

Nigel Lesmoir-Gordon has been an independent film and documentary maker since the 1960s.He is the co-author of Introducing Fractal Geometry, with Will Rood and Ralph Edney (2001), and is currently exploring the educational applications of fractal geometry

N. Lesmoir-Gordon (ed.), *The Colours of Infinity: The Beauty and Power of Fractals*, DOI 10.1007/978-1-84996-486-9_8, © Springer-Verlag London Limited 2010

Plato sought to explain nature with five regular solid forms.
Newton and Kepler bent Plato's circle into an ellipse. Modern
science analysed Plato's shapes into particles and waves, and
generalised the curves of Newton and Kepler to relative prob-
abilities – still without a single 'rough edge'. Now, more than
two thousand years after Plato, nearly three hundred years after
Newton, Benoît Mandelbrot has established a discovery that
ranks with the laws of regular motion.

Professor Eugene Stanley

In the late summer of 1991 I was reading Professor Ian Stewart's book on the new mathematics, *Does God Play Dice?* and Sir Roger Penrose's *The Emperor's New Mind*. On 12 August of that year a truly remarkable corn circle was discovered near Cambridge, whose formation has come to be known as 'The Ickleton Mandelbrot'. Ickleton is five miles from where I live. It quickly became part of local folklore. Reading Ian Stewart's and Roger Penrose's books I was simultaneously discovering for myself, for the first time, the Mandelbrot Set and fractal geometry. This was a revelation to me. I was reading the books as part of some ongoing research I was doing into cosmology and the new maths for a documentary I was hoping to make.

Interestingly in both books the chapters on the M-set have most poetic and evocative titles and both refer to the Set itself. Stewart calls his chapter 'The Gingerbread Man' and Penrose 'The Land of Tor' Bled-Nam'. In a mathematics book! Well, right off they looked intriguing, and indeed they were.

And once I had seen the Mandelbrot Set and then read about it, I was hooked. So what was it about the M-set that drew me in? It was that this mathematical entity should look so organic – like an insect or a hairy potato – and that it was infinitely complex and yet born from such humble beginnings – a simple equation with just three components. You would be hard-pressed to find anything simpler. As Professor Ian Stewart says in the film:

There's an interesting parallel with the equation that almost everybody is familiar with – the only equation that everybody is familiar with: $e = mc^2$. Albert Einstein's equation, which says that matter and energy are equivalent to each other. That was a very simple equation with very far-reaching consequences. And the equation for the Mandelbrot Set is equally simple: $Z \rightleftharpoons z^2 + c$

But there is one big, big difference, though, between the equation for the M-set and Einstein's famous equation. In Mandelbrot's equation there's no equal sign. Instead there's a double arrow. This works as a kind of two-way traffic sign, allowing the numbers to flow in both directions, constantly feeding back on themselves. The numbers go round and round in a loop. This effect is called iteration. The output of the first operation becomes the input of the second operation, becomes the output of the third operation and so on. In just this same way evolution proceeds in nature through an iterative feedback loop.

And the structures, which we find in the Mandelbrot Set do remind us of many things we

Professor Rudy Rucker wrote:

The image is a bit like a bug: a big warty buttocks-shape with a disk stuck onto it. There's an antenna sticking out of the disk, and shish-kababbed onto the antenna are tiny little Mandelbrot sets: buttocks, warts & disk. Each of the warts is a Mandelbrot disk, too, each with a wiggly antenna coming out, and with shish ka-babs of buttocks, warts & disks, with yet smaller antennae, buttocks, warts, and disks.

Now, I found it amazing that this extraordinary creature should be born from this ridiculously simple equation. Eureka! The power and beauty of math-ematics was revealed to me for the first time. I got it. The scales were lifted from my eyes and I could finally see the link between mathematics, the mind and the physical, observable universe.

The point is that the universe is described most effectively and accurately using the language of math-ematics. The subatomic world can only be described by physicists through the language of mathematics. There is no other way to talk about the very, very small.

There is one more extremely important aspect of this equation, which I have not yet mentioned, which is that z and c are complex numbers. Complex numbers are among the most important ideas in the whole of mathematics. Complex numbers have their own arithmetic, algebra and analysis. They rely for their existence on an act of pure mathematical imagination: that is, to agree that minus 1 is allowed

see in the natural world. There are spirals and ten-drils like the finest gossamer, each precisely formed. These curlicues look, not so much like crystal structures, as like plants, which have grown with a regularity amounting to perfection. Some resemble the tails of sea horses, others like spiralling seashells or the petals of some intricate flower. No matter how far we dive into the Mandelbrot Set, we can never reach the end of it, nor will it ever cease to create more patterns, more structures, more intricate forms. It is infinitely complex.

Once I had seen the Mandelbrot Set and then read about it, I was hooked. So what was it about the M-set that drew me in? It was that this mathematical entity should look so organic – like an insect or a hairy potato – and that it was infinitely complex and yet born from such humble beginnings – a simple equation with just three components. you would be hard-pressed to find anything more simple.

to have a square root – the square root of minus one: $\sqrt{-1}$. Complex numbers have two components or coordinates: 'real' and 'imaginary'. They were originally invented as a mathematical tool to help solve equations. For mathematicians now imaginary numbers are just as 'real' as real numbers. They are connected to reality at a deep level.

Since the square root of a negative number cannot be placed anywhere on the number line, mathematicians up to the nineteenth century could not ascribe any sense of reality to these quantities. The great Leibniz, inventor of the Differential Calculus, attributed $\sqrt{-1}$ a mystical quality, seeing it as a manifestation of the Divine Spirit. He called it 'that amphibian between being and not being'.

A century later, Leonhard Euler wrote in his work ALGEBRA in words that still echo the same sense of wonder:

All expressions such as the square root of minus one ($\sqrt{-1}$) are impossible or imaginary numbers, since they represent roots of negative quantities. Of such numbers we may truly assert that they are neither nothing, nor greater than nothing, nor less than nothing.

Carl Friedrich Gauss declared forcefully that 'an objective existence can be assigned to these imaginary beings'. Gauss realized that there was no room for imaginary numbers anywhere on the real number line, which runs from east to west. He took the bold step of placing them on a perpendicular axis, through the point zero and running from north to south. This creates a coordinate system, where all the real numbers are placed on the 'real axis' and all the imaginary numbers on the 'imaginary axis'.

Imaginary numbers! Wonderful! That was it. I was caught and completely snared in this wild, weird and wonderful world. And so it was that the Mandelbrot Set became the subject of the film I was to make. I saw it as a way into this mysterious mathematical world through which I could make this world accessible and fun for viewers.

I researched a lot more, wrote my treatment and submitted it to the BBC, Discovery and Channel 4. And, as night follows day, the almost predictable rejections followed. I fumed for a year, thrashing about for a way to finance the project. But the penny finally dropped and I came straight to the conclusion that what I needed was a name – a famous name or names, to attach to the project – some star quality to give it that special feel and appeal to financiers. A star yes, but it had to be the right kind of star for the subject – Maths! It was difficult.

There used to be a little shop in Ladbroke Grove called Strange Attractions. Sadly it's closed now, but at the time it was thriving, selling the products of the new mathematics – fractal geometry and chaos theory. Books, music, computer games, T-shirts, posters, coasters, cups and postcards. Inside, everywhere I looked, I saw the M-set.

Chatting with the sales assistant, I found out I had known the owner in London in the sixties. Among

Imaginary numbers! Wonderful! That was it.
I was caught and completely snared in this wild, weird and wonderful world.
And so it was that the Mandelbrot set became the subject of the film I was to make.
I saw it as a way into this mysterious mathematical world through which
I could make this world accessible and fun for viewers.

The crew with Nigel Lesmoir-Gordon and Arthur C Clarke in his study, from the left, standing: Editor & Designer, Simon Gilbert; Production Assistant, Artup Warnasiri; Production Manager, A K Warnarsiri; Producer, Paul Sinclair; Writer & Director, Nigel Lesmoir-Gordon; Seated: Director of Photography, John Lamborn and Arthur C Clarke

other things, Greg Sams pioneered the Vegiburger. Now it was chaos theory. Greg and I went out for coffee. We talked about old times, shared our excitement in the new maths and then I told him my problem.

Bingo! Two months before, Greg told me, Fred Clarke had been driving Sir Arthur C. Clarke back to his apartment in Holland Park, through Ladbroke Grove and past Greg's shop. Arthur C's sharp eyes caught a glimpse of the colourful snapshot of the M-set, swinging in the wind on a big sign over Greg's shop door. Arthur had Fred reverse the car back to the shop and park. Greg was in.

And now comes the good bit. Clarke had just returned from Riyadh, where he wrote that he had:

... the privilege of addressing the largest gathering of astronauts and cosmonauts ever assembled at one place (more than fifty, including Apollo 11's Buzz Aldrin and Mike Collins, and the first 'space-walker' Alexei Leonov) ... I decided to expand their horizons by introducing them to something really large! So, with astronaut Prince Sultan bin Salman bin Abdul Aziz in the chair, I delivered a lavishly illustrated

lecture: 'The Colours of Infinity – Exploring the Fractal Universe'.

Arthur gave Greg Sams permission to publish the speech. When I arrived Greg had just that week received the printer's proof. He gave me a copy and Sir Arthur's fax number in Colombo, Sri Lanka, and off I went to compose a persuasive letter.

I faxed my two-page plea the next morning, but forgot to inform my wife, Jenny, that I'd sent it, more or less expecting nothing to come back. So when Arthur C. Clarke phoned, all the way from Sri Lanka, two hours later, Jenny was completely unprepared. Only her quick responses saved the situation as she took on board what was happening. She calmed down, took notes and came away with his agreement to a two-day shoot at his house in Colombo.

Only one problem now: no money, love! But word got about surprisingly quickly and the finance started to flow. Not a torrent, but enough of a trickle for us to afford to take a British crew and equipment to Sri Lanka for four days.

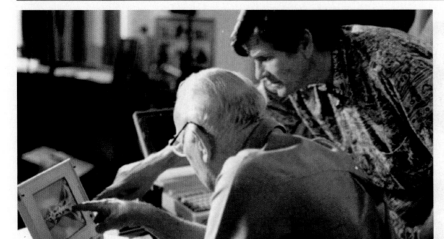

Jenny put in some of her savings, as did two of her friends, my close friend and my friend's Mum. The crew all agreed to forgo salaries and settle for percentage points according to the number of days/weeks (in my case years!) put into the project. Everything else had to be paid for: equipment, stock, flights, excess baggage, food, accommodation, insurances and so on.

Paul Sinclair bravely undertook to produce the programme, relying on me to cover the writing and directing. John Lamborn jumped at the chance of pointing his camera at Arthur C. Clarke.

Full of fear and trepidation, we entered Arthur C. Clarke's humble but high-tech apartment. I shook his hand and presented my offerings – three Mandelbrot mugs and some fractal coasters from Strange Attractions. As he was opening the package, he said, 'Nigel, if you hadn't turned up to make this film, I would have done it sooner or later with someone, somehow!' Nice timing.

Arthur and I talked for the best part of the next day, came to an agreement on the programme structure that I was proposing and wrote his links. We kept them all short. We had to. Autocue teleprompting was not within our budget!

Arthur C. Clarke was wonderful and passionate about the topic and furiously keen to do a good job. Fortunately, we saw eye to eye on the project immediately, and things stayed like that almost every inch of the way. In two days we had everything we needed in the can and were off.

By this time I had got hold of some animated colour pictures of the Mandelbrot Set on videotape. This tape had been made some two years earlier by a group of American mathematicians working in down time during the installation of the Cray supercomputer at Cornel University.

I needed somewhere to cut a short pilot together to show what we had done and give an idea of what still needed to be done. We were out of money. Boyd Catling of Original Films came to the rescue and offered me the use of his off-line suite when it wasn't booked out.

I'd known David Gilmour since teenage days in Cambridge and had watched with awe and wonder his rise to stardom with the Pink Floyd. So I took my pilot round to David's house, ran it for him a couple of times and, with very little persuasion on my part, he agreed to do the music for the programme.

Once again, it didn't take long for the finance to appear. But we needed a fair old wedge this time. We had to go to Warwick University to shoot Professor Ian Stewart, our star academic, and we needed to cross the pond too. Dr Benoît Mandelbrot, the discoverer of the set, was in upstate New York and our genius-inventor, Dr Michael Barnsley was at Iterated Systems in Atlanta, Georgia.

We managed eventually to pull all the finance we needed, and the same crew boarded the plane at Heathrow, only this time since we needed more help and took my daughter Daisy along as our PA. What struck me most forcefully throughout the shooting of the production was how happy all the contributors were to talk about the subject. And, I'd say, this aspect does come across loud and clear in the programme. I set out to make, not a deeply mathematical, analytical piece, but rather a celebration of a remarkable discovery. I wanted it to be fun. I knew it had to entertain at some level if it was going to reach a really wide audience.

With the shooting over, nothing remained but the edit. But how? Simon Gilbert, an old friend, offered his services. Simon and I then cast around for ways and means of off-lining. We couldn't go back to Boyd Catling again. In the event we put the programme together over a nine-month period in a whole host of different suites. most of it in a shed out by Heathrow Airport. Nights, weekends, holidays. You name it!

While Simon and I ached away on the low-band, a young mathematics graduate from Cambridge, Bill Rood, was working wonders on his Acorn Archimedes computer. We needed new fractal pictures, new M-set zooms and Bill could make them. Strange Attractions!

I can't say the off-line edit was all fun. It wasn't. It was painful. But we did get there, and getting there was just wonderful. I have to say the end product was all that I had hoped it would be. It was just as I imagined it would be. It was exactly to my original treatment. And, it had quite a bit more to it than I had ever expected. The thread was thrown beautifully from one participant to the next. The programme flowed and David Gilmour's music was a perfect fit.

The on-line was completed in fits and starts over an uncomfortably extended period at Essential Pictures and at Barrie Hinchcliffe Productions. Kevin Pyne did an expert sound dub and we had our finished 52-minute programme.

I sent a VHS copy to Arthur C. Clarke and he faxed back: 'I have now seen the programme a second time and am even more impressed. It really is stunning. I hope it wins lots of awards!'

That fax came as an enormous relief for me. I had expected him to like it, but nonetheless I really did breathe a huge sigh of relief when he responded so positively. Same thing again when I showed it to David Gilmour. He also thought it was excellent. I had the stamp of approval I needed from the big boys!

We had that, yes, and we had our programme, yes. But, we didn't have a distributor! No way for us to get it out there! No way to get it seen.

Fortunately, Paul Sinclair had been able to take the pilot cut down to MIPCOM in Cannes a couple of times. So seeds had been sown and we hoped that interest was growing. It had been. And, before we knew what had hit us, we were spoilt for choice. We had three potential players at our door. In the event we went with Beyond Distribution, based in Sydney. We signed our contract with them on 17 November 1995.

Since that date they have sold the programme for broadcast in Japan, Canada, Russia, Finland, Argentina, Venezuela, Russia, Poland, Hungary, Spain, Italy, the CIS, Israel, over 70 PBS stations in the USA, Thailand, Indonesia, New Zealand and Australia. 'Colours' was eventually broadcast in the UK by Channel Four on 7 September 1996.

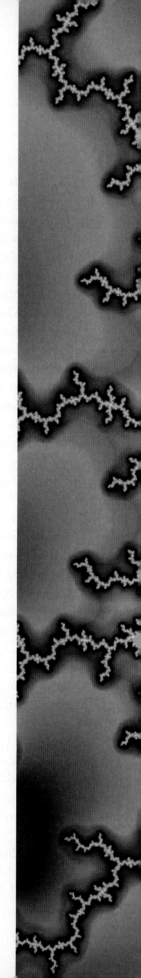

ARTHUR C. CLARKE Presents
THE COLOURS OF INFINITY-THE FILM SCRIPT

© GORDON FILMS UK 1995

'The most beautiful thing we can experience is the mysterious. It is the source of all true art and science.'

Albert Einstein

N. Lesmoir-Gordon (ed.), *The Colours of Infinity: The Beauty and Power of Fractals*
DOI 10.1007/978-1-84996-486-9_9, © Springer-Verlag London Limited 2010

FADE UP M-SET IN DISTANCE

M-SET ZOOMS UP SLOWLY UNTIL IT FILLS THE FRAME

ARTHUR C. CLARKE

This is the Mandelbrot Set: one of the most beautiful and remarkable discoveries in the entire history of mathematics. And yet it was discovered as recently as 1980.

MAIN TITLE FADES UP OVER M-SET WITH BURST OF COLOUR CYCLING:

'ARTHUR C. CLARKE PRESENTS THE COLOURS OF INFINITY'

M-SET AND TITLE MIX TO TRACK OVER PCB WITH M-SET FRACTAL SUPERIMPOSITION

ARTHUR C. CLARKE

The invention of the silicon chip in the 1970s created a revolution in computers and communication and hence transformed our way of life. We are now seeing another ... revolution which is going to change our view of the Universe and give us a better understanding of its working.

MIX FROM FRACTAL TO CRAB NEBULA

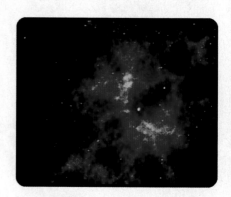

I'm Arthur C. Clarke. I write Science Fact and Science Fiction. You may know my movie, 2001– A Space Odyssey. I've seen some remarkable developments and inventions in my lifetime, but one of the most extraordinary is the Mandelbrot Set and Fractal Geometry. This film will explore the Fractal Universe. And on our voyage of discovery we will be helped by ...

PIC OF IAN STEWART

Professor Ian Stewart of the Mathematics Institute, University of Warwick and author of over 100 published scientific works.

PIC OF MICHAEL BARNSLEY

Dr Michael Barnsley, former Professor of Mathematics at Georgia Institute of Technology who received a \$$2\frac{1}{2}$ million government grant in 1991 to develop Fractal Image Compression Systems.

PIC OF STEPHEN HAWKING

Professor Stephen Hawking, the Mathematician and Cosmologist and author of the bestselling book A Brief History of Time.

PIC OF BENOÎT MANDELBROT

And finally Dr Benoît Mandelbrot whose unorthodox mathematics led to the discovery of the Mandelbrot Set and Fractal Geometry.

IAN STEWART

(Prof. Ian Stewart Mathematics Institute, University of Warwick)
I first saw the Mandelbrot Set somewhere in the mid-eighties. I remember it quite clearly. We were at a mathematical conference on something totally different. And everyone went along to this exhibition because it was mathematical pictures. And there were these amazing coloured pictures on the wall. And I'd really not seen anything like this before.

M-SET ON BLACK, ROTATING THROUGH 180 DEGREES

It's not easy to describe the Mandelbrot Set visually. It looks like a man. It looks like a cat. It looks like a cactus. It looks like a cockroach. It's got little bits and pieces that remind us of almost anything you can see out in the real world. Particularly living things. So it has a character that reminds us of a lot of things. And yet it itself is unique and new.

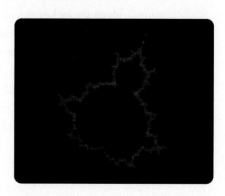

MICHAEL BARNSLEY

(Dr Michael Barnsley Chief Scientist, Iterated Systems Inc.)
The Mandelbrot Set is real. An absolute thing – no question whatsoever! Any mathematician, or any computer scientist, or student in a school can study it and find the same – describe the same – thing. It's a common experience.

M-SET ON BLACK WITH ZOOM INTO BOUNDARY

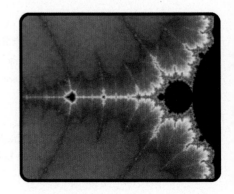

And so such things that can be magnified forever, and have infinite precision, do exist. But, they're not touchable.

IAN STEWART

It's a geometrical shape, an icon if you wish, which somehow embodies as an example, a very important aspect of how the world works. Somebody recently actually called this set 'The Thumbprint of God'!

ARTHUR. C. CLARKE

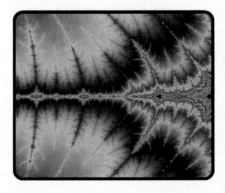

Now we'll begin our serious exploration of the Mandelbrot Set. A voyage, which in fact, could last for ever and ever – much longer than the lifetime of the Universe! I have here the full Set – about six inches across. Now, if I blow this up: I'll increase the magnification thirteen times. And you'll see more and more detail is appearing. And the interesting thing is you see mini-Mandelbrots-replicas – almost identical, yet perhaps, subtly different from the original Set. And I can go on doing this. Here's a magnification of more than three thousand times. So, the original picture – about six inches across – is now half a mile across! And no matter how much we magnified it, a million times, a billion times – until the original set was bigger than the entire Universe – we would still see new patterns, new images emerging, because the frontier of the M-set is infinitely complex. And when I say infinitely, I really mean that. Most people when they say infinitely, they mean – oh – only, rather big. But, this is really infinity!

M-SET ZOOM SEQUENCE

What is so remarkable, in fact astounding, about the Mandelbrot Set is that, although it is ... infinitely complex, it's based on incredibly simple principles – unlike almost everything in modern Mathematics. In fact, anybody who can add and multiply, can understand the principles on which it's based. You don't even have to subtract or divide, still less use logarithms, or trigonometrical functions to comprehend how the Mandelbrot Set is created. In fact, in principle, it could have been discovered any time in human history, and not merely in 1980. But the problem was this, although it's only based on adding and multiplying, you have to carry out those operations millions, billions of times, to create a complete Set. And that's why it was not discovered until the era of modern computers.

EXTERIOR: THOMAS J. WATSON RESEARCH CENTER, NEW YORK

It was on the 1st March 1980 at IBM's Thomas J. Watson Research Centre in upstate New York, that Benoît Mandelbrot first glimpsed the M-set.

B&W STILL BENOIT MANDELBROT: STILL OF FIRST PRINT OF M-SET

The seeds of this discovery were in fact sown decades before the M-set was first seen. In Paris, in 1917, a mathematician called Gaston Julia published papers connected with so-called complex numbers. The results of his endeavours eventually became known as Julia Sets. Although Julia himself never saw a Julia Set! He could only guess at them. And it wouldn't be until the advent of modern computers that Julia Sets could be seen for the first time.

STILL OF GASTON JULIA: B&W STILL OF JULIA SET

BENOÎT B. MANDELBROT

For me the first step almost with any difficult mathematical problem was to program it, and see how it looked like. We started programming Julia Sets of all kinds. It was extraordinary great fun! And in particular, at one point, we became interested in the Julia Set of the simplest possible transformation …– Z goes to Z squared plus C. So Z times Z plus C.

MONTAGE OF FIVE B&W STILLS OF JULIA SETS

STILL OF ROUGH PRINT OUT OF M-SET

> I made many pictures of it. First of all, the first one was very rough.

THREE STILLS OF FIRST M-SET PRINT-OUT

> But the very rough pictures, they were not the answer.
> Each rough picture asked a question. So I made another
> picture, another picture. And after a few weeks we had
> this very strong, overwhelming impression that this was a
> kind of big bear we have encountered.

ZOOM IN ON M-SET

> I think the most important implication is that, from very
> simple formulas you can get very complicated results. It's
> fundamental from the viewpoint of the very basis of sci-
> ence. Because, what is science? We have all this mess
> around us. Things are totally incomprehensible. And then eventually – more
> or less rapidly, more or less hard to achieve– we find simple laws, simple
> formulas. In a way, a very simple formula Newton's Law,
> which is just also a few symbols can, by hard work, explain
> the motion of the planets around the sun and many, many
> other things to the 50th decimal! It's marvellous: a very
> simple formula explains all these very complicated things.

*SOLAR SYSTEM GRAPHIC WITH ZOOM OUT AND CAPTION OF NEWTON'S
EQUATION SUPERIMPOSED*

IAN STEWART

> There's an interesting parallel with the equation that almost everybody is
> familiar with – the only equation that everybody is familiar with – $E = mc^2$.
> Albert Einstein's equation, that says matter and energy are equivalent to
> each other.

EINSTEIN STILL WITH EQUATION SUPERIMPOSED

That was a very simple equation with very far-reaching consequences. And the equation for the Mandelbrot Set is equally simple. $Z = Z^2 + c$.

M-SET GRAPHIC WITH MANDELBROT'S EQUATION.

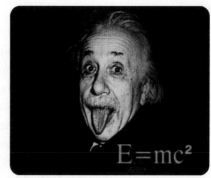

ARTHUR C. CLARKE

The letters in the Mandelbrot equation stand for numbers, unlike those in Einstein's equation, where they stand for physical quantities – mass, velocity, energy. The Mandelbrot numbers are coordinates, positions on the plane, defining the location of a spot.

M-SET GRAPHIC WITH MANDELBROT'S EQUATION

Another difference from Einstein's equation – and a very important one – is this double arrow. It's a kind of two-way traffic sign. The numbers flow in both directions, constantly feeding back on themselves. This process of going round and round a loop is called iteration. It's rather like a dog chasing its own tail: the output of one operation becomes the input of the other, and so on and on.

M-SET GRAPHIC WITH ITERATING NUMBERS

When the Mandelbrot equation is given a number representing a point, and that number is iterated through the equation, one or two things happen. Either the number gets bigger and bigger and shoots off to infinity. Or it shrinks to zero. Depending on which happens, the computer then knows where to draw a boundary line. So, what we get from this basic iteration is a kind of map, dividing this world into two distinct territories. Outside it are all the numbers that have the freedom of infinity. Inside it, numbers that are prisoners, trapped and doomed to ultimate extinction.

M-SET GRAPHIC WITH ZOOM INTO BOUNDARY

IAN STEWART

Think of a computer screen. You're looking at each individual little element, each pixel of the screen. You pick one of these pixels. You apply this rule lots and lots of times, and either the pixel moves off and disappears completely from view, or it moves in towards a fixed point in the middle of the screen. And what you do is, you just want to distinguish between going off out to infinity or going into zero. So any point that moves into zero when you apply this rule, you colour that point black. And any point that goes off to infinity – what people tend to do is colour it all sorts of wonderful, rainbow hues about how fast it goes away. The important bit is the black bit in the middle that's all the stuff that doesn't escape when you keep applying this rule.

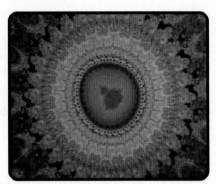

ARTHUR. C. CLARKE

Now the colours are completely arbitrary. They could be anything. But they are not meaningless. A very good analogy is the contour map you'll see of a mountain range for example, where the contours are drawn and coloured – the areas are coloured. The highest areas might be coloured white, then brown, then green, and then if you went on into the sea, deeper and deeper blues – just to show where the various levels occur. So, it's the same here. You can make the colours anything you like, but they do define the different areas of calculation. And you can change them and get the most gorgeous results! Just look at this.

COMPUTER SCREEN SHOWING INTERIOR OF M-SET WITH COLOUR CYCLING

Now, you may think that the frontiers are moving, but there's no motion whatsoever. Only the colours are cycling. In or out. Nothing is moving.

COMPUTER SCREEN SHOWING M-SET WITH COLOUR CYCLING

BENOÎT MANDELBROT

When you get very, very fine to very, very small details, the variations can become of overwhelming complexity. So complex that no single picture can possibly give justice to them. It's impossible, in black and white or coloured picture, to show how complicated it is. The only way is to use what we call a 'colour cycle'.

COMPUTER SCREEN SHOWING M-SET WITH COLOUR CYCLING

That is, the colours change regularly and each set of colours, in a way, reveals a different property of these variations – a different variation. So by having this colour cycling, one reveals in a very, very strong fashion the extraordinary complexity of the Set. If the whole Set were represented this scale, the end of the Set would go so far as to go somewhere near the star Sirius! Very, very far. Enormously big! For this very tiny speck. Yet, in the middle of the speck you see an exact replica of the whole.

DEEP ZOOM INTO M-SET WITH COLOURS CYCLING

IAN STEWART

One of the most striking facets of the Mandelbrot Set is the internal consistency of the object. It all hangs together. And if you look at the boundary and zoom in – if you look in just the right place, what you see is baby Mandelbrot Sets. Perfect in every detail. They're just slightly bent compared to the real Set – you can't even see that. But if you look closer you see they are. And they're decorated by slightly different external features. And then, by the same token, if you zoom into the boundary of those, you see baby, baby Mandelbrot Sets! The second generation. And inside those the third generation. It goes on forever. And so, you're seeing islands of order in a sea of chaos.

ARTHUR C. CLARKE

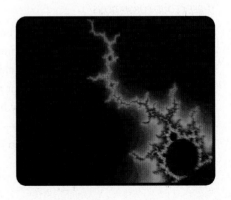

I'm sure it's occurred to you that the Mandelbrot Set looks like some kind of strange insect. It certainly has an organic feeling about it. It's got warts all over it and it's also quite hairy!

ZOOM INTO M-SET DENDRITE

If you go out along one of these hairs we find something rather interesting. Now, look what's happening: at the tip of each hair it splits into two others. And so on. Each is splitting, going on indefinitely. This splitting up – this 'bifurcation' – going off into apparently random directions quite abruptly, is typical of a class of mathematical entities called Fractals. The Mandelbrot Set is the most famous Fractal.

IAN STEWART

The word Fractal means any geometrical structure that has detail on all scales of magnification. No matter how big you make it, you still see extra, new details you didn't see before. And the name was actually invented by Mandelbrot himself. He felt he had to have a name for this area he realized he was working in. And so he coined the word Fractal because it conveys this feeling of fragmented, broken, fractional, irregular.

M-SET STILL

MICHAEL BARNSLEY

Chief Scientist Iterated Systems Inc.

It took a long time for us to emerge and start to look out at the other part of the physical, observable Universe. Not as narrow, studied little entities – the scientist, who studies the flea on the back of the flea on the back of the flea. But rather being able suddenly to look out at the totality of nature, and then say, 'my goodness me, we've got nothing to describe this with!'. Clouds are not made with straight edges. Trees are not circles, they're not triangles, they're something very, very different indeed.

THREE CLOUD FORMATIONS

But there is a continual kind of a pattern that I can see as I look at the edge of a rising cumulus cloud, one of those very, very wrinkly, coruscated clouds that has such fine structure in it. And you say, but there's no lines or circles there. The wonderful discovery has been that there's an extension of classical geometry – Euclidean Geometry – which is called Fractal Geometry.

BENOÎT MANDELBROT

Fractals are shapes which we are extraordinarily used to in … how to say … in our subconscious, ill-organized life. For example, everybody knows that if you take a map of Britain on a small school globe, you see a very simplified shape. Cornwall is just a kind of triangle and Wales perhaps a little rectangle. You cannot put the detail on a small map. If you look at a larger map, you add more detail.

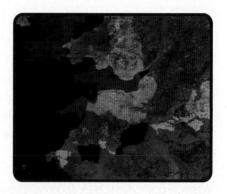

SLOW ZOOM IN THROUGH THREE SUCCESSIVE MIXES ON GRAPHIC OF PLANET EARTH

The closer you come, in a certain sense – imagine yourself like somebody coming in a rocket: from far away you see very little. The closer you come, the more detail you see. If you come very, very close you begin to see rocks.

WAVES BREAKING OVER ROCKS

And finally the idea of coastline disappears, because one doesn't know any longer where is land and where is water. So a need arose in my mind to put together a geometry, based upon many known facts in mathematics – scattered facts in mathematics – many scattered facts in our experience, many scattered facts in the results of what scientists had done of various kinds putting together all of these things, and using them as bricks, if you will, of a new building, which is a new geometry, which is a geometry of shapes, which are equally rough at all scales.

M-SET DETAIL WITH COLOUR CYCLING

ARTHUR C. CLARKE

One of the revolutions in thought that's resulted from this discovery is the realization that nature deals not in the smooth, continuous objects, as we always imagined, but more often in Fractals. And I'd like to show you how she does this.

INTERIOR: CLOSE SHOT COMPUTER TERMINAL WITH FRACTAL PROGRAM

Now I'm going to generate a Fractal before your very eyes!

INTERIOR: COMPUTER SCREEN, SHOWING FRACTAL 'SEED'

What you see here is what is called the 'seed' and it's an appropriate word in this case. Those two lines represent the first generation of the formation of a geometric figure. And the computer has been told to continue growing these lines, but changing the direction every so often and at different distances. Now, that's a very simple set of instructions.

INTERIOR: COMPUTER SCREEN, SHOWING GROWTH OF FIRST FRACTAL TREE

But look what happens after they have been carried out for, say, ten generations. The tree I showed here – and it does look very much like a tree in nature – is symmetrical because the two branches at the beginning were the same length, off in the same direction. But if we change the length of one branch and change the direction, look what happens.

INTERIOR: COMPUTER SCREEN SHOWING GROWTH OF SECOND FRACTAL TREE

In a way this is a more realistic tree than the first one, because in nature you seldom have perfect symmetry.

COMPUTER-GENERATED OBJECTS: FERN; SEA-SHELL; MOUNTAINS

Much more elaborate structures can be created by very similar rules. I would like to emphasize that all of these shapes, or objects, or whatever you like to call them, although they look real, are generated entirely in the computer by following out a few simple instructions, and repeating them over and over again. This is the way in which nature creates things.

BUTTERFLY'S LIFE CYCLE: STARTING WITH GRAPHIC OF COMPLEX MOLECULE, THROUGH TO EMERGENCE OF THE BUTTERFLY

MICHAEL BARNSLEY

It's exactly like the DNA in a butterfly's egg. Somehow that unravels and unrolls to form the extraordinary and beautiful pattern on a butterfly's wing with its myriads of colours and form. Somehow it's hidden in that seed, in the DNA. And not only that, but the wings themselves probably only occupy a relatively small part of the total DNA. They are, if you like a little formula, that is unravelled by the process of growth and deterministic following of rules to form this natural and beautiful thing.

FRACTAL IMAGE WITH COLOUR CYCLING

FRACTAL IMAGE WITH COLOUR CYCLING AND BUTTERFLIES SUPERIMPOSED

IAN STEWART

Living creatures seem to be complicated structures produced from simple rules, simple laws of physics and chemistry. And a lot of the structure that you see in living creatures – this organic, but patterned structure – leaves on trees, ferns particularly things like that – have the same feature that the Mandelbrot Set has.

CLOSE SHOT FERN LEAF WITH RIVER IN BACKGROUND

You look at little pieces of them and they have lots and lots of detail. And in fact the little pieces look very similar sometimes to the whole thing. It's very tempting to compare the way a simple formula produces the complicated Mandelbrot Set with the way very tiny things in nature produce complicated organisms. And there are certainly some similarities, in that there is the same kind of unfolding of a process. The instructions are there but not an actual description of the object.

WIDER SHOT OF RIVERBANK WITH FERN

TRACK ALONG M-SET LEFT TO RIGHT

MONTAGE OF TREES, FOLIAGE, BRANCHES, FLOWERS

MICHAEL BARNSLEY

Once you've developed a Fractal Geometer's eye, you can't help but see them everywhere. Every single thing you see is, one way or another, described by reference either to itself or to something else in the picture you see. It's as though you're staring at a vast dictionary but the dictionary words are bits of pictures and the references, the definitions of the words, are made with other bits of pictures. So you stare at one picture.

I look out in the garden and at the trees, and I see this set of relationships between the picture and other bits of the picture.

GRAPHIC OF BLOCK MOVING ON SCREEN

Those relationships are no more or less than the assertion – from my point of view – that what I'm seeing is Fractals everywhere!

ZOOM IN ON FOLIAGE, FOLLOWED BY TRACK ALONG M-SET FROM LEFT TO RIGHT WITH COLOUR CYCLING AND WITH LEAVES, TREES VINES DISSOLVING IN AND OUT

IAN STEWART

The discovery of Fractal Geometry changes completely the kind of patterns we can look for in nature. And that is really a fundamental change to the sort of things mathematicians and scientists can do. And that's got to have a big effect.

M-SET WITH COLOUR CYCLING AND CLOUDS SUPERIMPOSED

Fractal Geometry is already being applied throughout the physical sciences as a way of describing data in a new way. And the dream is that a Fractal Geometer can describe a cloud as simply as an architect can describe a house! He can use his intricate, repeatable formulas – simple formulas – to describe these unimaginably complex and beautiful shapes, and then communicate them: from me to another scientist, to you. Here's, not my straight line, build it straight, but here's my ragged formula – but it's very simple. Go build it wild like this!

ZOOM OUT FROM RAIN FOREST RIVER

MICHAEL BARNSLEY

Can sort of think there might even be the sort of semaphore of nature – of the physical world – of how it tells itself what it's supposed to be.

TRACK ALONG TOP OF M-SET SPIKE FROM LEFT TO RIGHT

ARTHUR C. CLARKE

Let's go back to the Mandelbrot Set and look at some
more of the strange flora and fauna of the Mandelbrot zoo.
There's a certain similarity between these shapes. We can
recognize they are cousins of each other, and yet they're
all different, despite their similarity. There's an infinite
variety here, just indeed as there is in the world of nature.
We see shapes that remind us of Elephants' trunks, tentacles
of octopae, sea horses, compound insect eyes. There's some
connection between the Mandelbrot Set and the way
nature operates!

FRACTAL WITH COLOUR CYCLING MIXES TO TILT FROM SPIRAL GALAXY

ARTHUR C. CLARKE

ARTHUR C. CLARKE ON ROOF WITH TELESCOPE: STILL OF THE PLANET SATURN

I'm looking at Saturn, one of the most beautiful objects in
the sky. In fact we've discovered quite recently that the
beautiful rings of Saturn, which have intrigued astrono-
mers for centuries, do illustrate some of the phenomena
we've been discussing in the Mandelbrot Set.

ZOOM IN ON STILL OF SATURN'S RINGS

As you go closer and closer at Saturn, you see more and more detail, which
no one had ever dreamed of before the space age opened and we were able
to get close-ups of Saturn and its rings. It's not surprising that when we
have so many examples of Fractals and related phenomena here on this
planet, that there are even more in the heavens.

SLOW ZOOM IN ON STILL OF MILKY WAY GALAXY

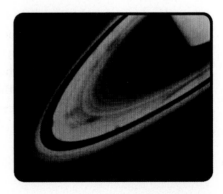

MICHAEL BARNSLEY

To me just looking up at the Milky Way is staring at a Fractal.
It's got an extraordinary dotty character and yet, if you take a
magnifying glass to it – that is, a telescope – and you look at
it ever closer you find that there are hundreds and thousands
more little dots where you thought there were almost none.

ZOOM IN ON MILKY WAY GALAXY

So you get an immediate example of a structure that seems to go in and in and in with more and more detail.

ARTHUR C. CLARKE

I had the great privilege of having a discussion with the famous cosmologist, Stephen Hawking, when I passed through London recently. And I said him, 'Dr. Hawking, the Mandelbrot Set is infinite in detail. You can explore it forever and ever – zoom into it. The real universe, however, does seem to have limits. As you go down into the micro-world you get, of course, molecules, atoms, neutrons and subatomic particles – quarks. But, does the real universe go on forever, is there a limit – a basement – unlike the Mandelbrot Set?'

STEPHEN HAWKING

Prof. Stephen Hawking Lucasian Professor of Mathematics Cambridge University

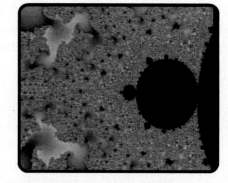

In the case of the universe there seems to be a limiting scale. It is called the Planck Length, and is about a million, billion, billion times smaller than an inch. This means that there is a limit to how complex the universe can be. It also means that the universe could be described by a theory that is fairly simple, at least on scales of the Planck Length. I just hope that we are smart enough to find it!

ARTHUR C. CLARKE

He thinks that there is a limit in the real universe: there's a small size below which nothing exists, called the Planck Length, which is about a million, million, million, millionth of a centimetre. Unimaginably small! But that is the fundamental unit of size, the sort of grittiness of the universe – nothing smaller than that. So perhaps the real universe does end there in smallness. But we're not sure. It may indeed go on forever like the Mandelbrot Set. We just don't know yet.

TRACK OVER M-SET RIGHT TO LEFT

STILL OF MICHAEL FARADAY DEMONSTRATING IN LABORATORY

I'm often asked, well these pictures are very pretty, but what's their practical value? And I'm tempted to answer in the famous words of Faraday, who once said, when someone asked him what use were these experiments – playing with wires and magnets – 'What use is a new-born baby? Faraday is also supposed to have told the Prime Minister, 'One day, Mr Prime Minister, you'll be able to tax it!' And in fact Fractal Geometry, the sort of things we have been demonstrating, has enormous commercial value.

IAN STEWART

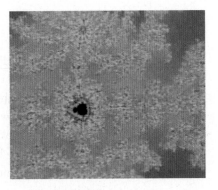

I think the discovery of the Mandelbrot Set and of Fractals in general is very important. It's important at the moment on an intellectual level more than hard-core technological level. There are some applications, but it's not yet put an important new gadget into every home, whereas things like the silicon chip certainly have. So most mathematical developments are like this. The ideas must come first, and then you have to translate them into practical things. And you can already see the beginnings of that translation occurring.

MICHAEL BARNSLEY

No longer do you have to draw a straight line through your data to make science of it. You can actually draw some Fractal Curve through it or measure some Fractal Dimension of the data and do science! So the first application is in terms of a better description of the physical observable world. There's a new branch of mathematics available to all scientists, and that application will stretch on through the centuries now as the primary tool for descriptive physical science.

BENOÎT MANDELBROT

Phenomena of great irregularity are very, very widespread in nature. In the study of what's called Condensed Matter – Polymers – such physical problems – one finds shapes of extremely great complication. These shapes could not be examined as geometric shapes before because there was no language to describe them. One couldn't describe the shape. One could only say things indirectly about them. Say, if you make such and such a measurement on them you'll get such and such a result. But that is, in a certain sense a shadow of the object. It's the effect the object had on a certain measurement procedure. But the object itself could not be described with the other geometry.

COMPLEX MOLECULE ROTATING WITH COLOUR CYCLING FRACTAL BACKGROUND

That's a very mundane example, but it's just the tip of the iceberg. There're an enormous number of structures, which are indeed only describable in terms of Fractal Geometry.

MICHAEL BARNSLEY

So the primary application will be as a tool for science in its own right. Science, and then engineering, and on through into the building of the next generation of devices and equipment that will follow from that in terms of the sort of application that we think of. You know, will there be a new type of – not computer – because before you perceived, understood, about desktop computers they weren't here, one didn't imagine them.

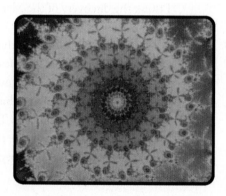

M-SET SPIRAL ZOOM WITH COLOUR CYCLING

But there will be new devices, extraordinary new devices based on the principles of Fractal Geometry, that will emerge over the next centuries.

ARTHUR C. CLARKE

Suppose you were the owner of a television station, or a satellite, which could broadcast just one television programme, and somebody came to you and said, 'with the same amount of power you can broadcast not one but ten programmes! What would that be worth to you?' Obviously it would multiply the value of your investment ten times overnight! Well, that's the sort of thing that Fractal Geometry makes possible.

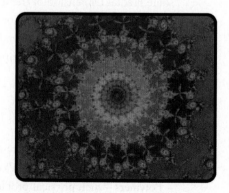

EXTERIOR: ATLANTA SKYLINE, EFFECTS INTO M-SET, WHICH RESOLVES INTO 3-D IMAGE

Atlanta, Georgia, headquarters of Michael Barnsley's Iterated Systems. In 1991 Barnsley received a $2\frac{1}{2}$ million government grant to develop fractal image compression systems. Corporations such as Microsoft, Mitsubishi, Multicom and Virgin, now use Barnsley's Image and Data Compression software.

INTERIOR: ATLANTA: MICHAEL BARNSLEY WITH DESIGN TEAM IN LONG SHOT

MICHAEL BARNSLEY

One of the most exciting moments occurred when I discovered the Collage Theorem. We'd been trying to work out how you could control a certain class of dynamical systems to make pictures of leaves.

COMPUTER SCREEN WITH FRACTAL STRUCTURE

Then, struggling with the question, it just dawned on me – that it was very simple. You needed to form a collage: a covering of the object with copies of itself – smaller shrunken copies. So that the whole object was tiled with copies of itself. It's a self-reference statement. It's as though you took a – you might take a triangle and cover it with little triangles. You might take a square and cover it with little squares.

COMPUTER SCREEN WITH THREE FERN TRANSFORMATIONS

Well, the theorem said, if you took a fern and covered it with little ferns, then you would have created a dynamical system or a formula for a fern. But, if you tried to actually create a picture using the Collage Theorem, it took hundreds of hours of graduate student time, working on the problem. And the Holy Grail at this point for us becamethe question of could we find a way to: tell a computer, just look at a picture – a digital picture – and automatically go ahead and find the fractal formula for it.

FRACTAL WITH COLOURS CYCLING

The discovery of how to automatically calculate the collage for an arbitrary picture came to me in a dream. From the early days of doing mathematics I used to have a recurrent nightmare, which was something to do with studying matrixes – may kind of remind one of, perhaps, an old-fashioned telephone switchboard. Well, in the dream what happened was there were thousands of holes and lots of wires connecting everywhere to everywhere. And it was always a sort of tense muddle between the switchboard with all the wires going everywhere, always in a horrible tangle. And somehow it represented a matrix.

SUPERIMPOSED IMAGES OF CIRCUIT BOARDS, WIRING CHANNELS, SWITCHBOARD WIRING

FRACTAL SURROUNDING MICHAEL BARNSLEY. IN OFFICE

I'd had this nightmare many, many times over twenty years. The night of the anniversary of my father's death – two years after – suddenly I saw in the dream how you could straighten out the switchboard, how all the wires would become untangled and be nicely connected.

FRACTAL WITH COLOUR CYCLING, EFFECTING INTO SPLIT SCREEN

And how you would join all the wires from big blocks to little blocks in the grid. And I woke up in the morning and knew that I'd discovered it. This was the total secret to fractal image compression: how to automatically look at a digital picture – these ones made of the … low resolution input, like your eye receives – and how to turn it into (a) a formula and (b) an entity of infinite resolution.

FRACTAL WITH COLOUR CYCLING

So the goal is now to be able to capture this Fire of Prometheus, if you like, this fractal wonder, put it in a box and being able to make this available to everyone.

ARTHUR C. CLARKE

ARTHUR C. CLARKE IN GARDEN WITH COMPUTER TERMINAL

Using Michael Barnsley's system we can now compress images so they require vastly fewer bits to store them. Now, whether you compress or expand an image depends on the same fractal principles, and Barnsley's theories have now become a commercial reality. Let's take a look at one of his programs. If you take a tiny piece of this image, which has been stored in digital form in the normal way and blow it up ... it becomes very pixilated. And they're huge pixels!

COMPUTER SCREEN WITH PARROT'S EYE

Now, if we take this very coarse image and pass it through Barnsley's Fractal Analyser we can actually reconstruct the details of the original image. If we then put the two images side by side the difference is startling.

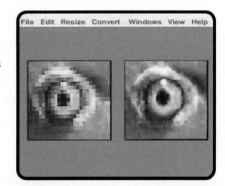

PIXELATED AND FRACTAL IMAGE SIDE BY SIDE ON SCREEN

So, where has all this detail come from? Well, this fractal image is a prediction, based on the digital data sample at the original low resolution.

FULL SCREEN MAGNIFICATION OF FRACTAL IMAGE

And you can of course magnify this image just as much as you wish. Because, like the M-set, it has infinite resolution.

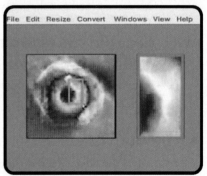

MICHAEL BARNSLEY

What you see are fractal textures – fractal creations – that mimic the missing data. They are, if you like, interpolations or predictions, but they are done using Fractal Geometry. What happens is, the original data is modelled by a fractal formula, and then we are looking at that Fractal in greater and greater detail.

FURTHER EXPANDED FRACTAL IMAGE OF THE PARROT'S EYE

ARTHUR C. CLARKE

We are all used to seeing, every night on TV, these satellite views of the Earth, with clouds moving over continents, showing the formation of storms and so forth. Now, these weather satellites have been operating for decades. What is not so well known is that there are also satellites up there – so-called spy satellites or reconnaissance satellites – which produce images of the Earth or, at least, points of particular interest to the military – with thousands of times the definition of the weather satellites! This means that they have to transmit tremendous amounts of data to the ground. Far more information than the weather satellites.

ANIMATION OF SATELLITE ORBITING THE EARTH

So, therefore, data compression – the ability to squeeze images, and send them, and then expand them again on the ground – is of enormous importance to the military. And we can thank those satellites for the fact that World War III has not yet broken out, and – hopefully – never will break out.

MODEL OF HUMAN CIRCULATORY SYSTEM

Fractal Geometry has surprising applications in medicine. This is the blood circulatory system of the human body.

MODEL OF BLOOD CIRCULATORY SYSTEM OF THE HUMAN HEAD

And yet, you'll recognize it. It is a kind of Fractal. Now we can understand what is really happening when our blood circulates.

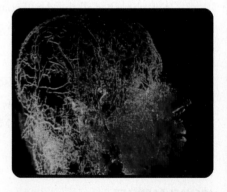

GRAPHIC OF BRAIN CIRCUITS

Here is the most important Fractal of all in the human body: a small portion of the incredibly complex wiring circuit of the brain. We may never understand how our brains work, but if we do, I suspect they will depend on some application of Fractal Geometry.

M-SET ZOOM

Why I think there may be some connection between the Mandelbrot Set and the wiring of the brain is because when I close my eyes – press my fingers against my eyelids, I see these patterns.

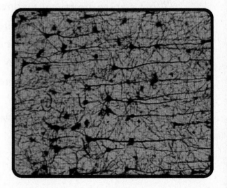

M-SET ZOOM

I'm sure you are all familiar with them. You also see them when anybody gives you a bang on the head! Sometimes these patterns echo some of the shapes of the Mandelbrot Set. Also I am told – I have never tried this experiment myself – but when certain illegal chemicals are ingested, you experience visual hallucinations strikingly similar to some of the patterns of the Mandelbrot Set. Why do these strange patterns have such an appeal? Well, obviously they trigger some kind of resonance in the mind. And, incidentally, there's an odd coincidence here: the name 'Mandelbrot' and the word 'Mandala' for a religious symbol. I'm sure it's a pure coincidence but, the Mandelbrot Set does indeed seem to contain an enormous number of Mandalas or symbols.

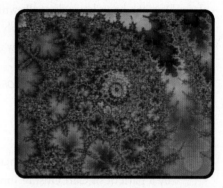

TIBETAN MANDALAS MIXING INTO M-SET

PAISLEY PATTERN MIXING INTO M-SET

The Paisley Pattern is one, and I am sure there are many others.

STAINED GLASS WINDOW

And in ecclesiastical design – such as stained glass windows, particularly in Islamic art, we find many echoes of the Mandelbrot Set centuries before it was discovered!

CARPET AND RUG DESIGNS

BENOÎT MANDELBROT

I had this experience, which many people repeated and told me about. I had this experience immediately: that when I first saw them, I was the first person to see them! There was absolutely no way anybody could have seen before. Yet, after a few days, or sometimes a few hours, a few minutes, it became almost familiar. I was finding features in it, which I have seen somewhere. So where I have I seen them? Well, first of all certainly, as I've said, in natural phenomena, but also, perhaps, in art.

CARPET MIX TO M-SET WITH COLOUR CYCLING, THEN WITH MODEL OF HUMAN BRAIN MIXED INTO THE CENTRE OF THE M-SET, AND OUT

So, one wonders, why is it so? We know the brain has some cells, which handle shapes, boundaries, and other cells, which handle the colour. Does the brain have also cells which handle Fractal complication? Well, we don't know. It's a purely hypothetical question. It's a tempting question, but we don't know anything about it.

MONTAGE OF FOUR PAINTINGS BY A PATIENT OF C.G. JUNG

ARTHUR C. CLARKE

Here's another strange resonance. This series of paintings was made in 1928 by a patient of Carl Gustav Jung, the co-founder of modern psychology.

STILL OF C.G. JUNG

Jung would have been surprised and delighted to know that the computer revolution, whose beginnings he just lived to see would give new impetus to his theory of the Collective Unconscious.

STILL OF PAINTING BY A PATIENT OF C. G. JUNG

MONTAGE OF FOUR M-SETS WITH COLOURS CYCLING

The idea that there is a well of consciousness, compounded of primordial, universal images that we all share. The substructure, or background of awareness. The mind clearly finds resonances in the M-set. But there are other, wider, implications too. This mathematics offers new insights into the way the universe works.

ZOOM OUT OF STILL OF SPIRAL GALAXY

How much in life is determined, and how much is due to chance?

ZOOM IN ON STILL OF SIR ISAAC NEWTON

IAN STEWART

When Isaac Newton came up with laws of motion and laws of gravity the picture that emerged was of a clockwork universe.

ZOOM OUT OF STILL OF SPIRAL GALAXY

It was of a machine that ticked on a predetermined course. All we needed to know was where it was now and what it was doing now, and then you could predict the future forever. And there are two challenges to this. One is Quantum Mechanics, which says that in fact there is irreducible chance built into the very fabric of the universe. And you can't actually say exactly what it's doing now. You can't say exactly what it's doing ever. But the other is, things that come out of the Mandelbrot Set and related parts of mathematics, which says that even in a Newtonian world, in practice you may not be able to predict the future. It can be deterministic in principle, but not in practice.

MICHAEL BARNSLEY

This is how God created a system, which gave us free will. It's the most brilliant manoeuvre in the universe: to create something, in which everything is free! How could you do that?

M-SET WITH COLOURS CYCLING

GRAPHIC: EINSTEIN AND DICE

IAN STEWART

Albert Einstein refused to accept the idea of a dice-playing deity. He wrote a letter to Max Born in which he said, 'you believe in a God who plays dice, and I in complete law and order'. So he obviously felt that chance and deterministic laws were not compatible. And he preferred the deterministic laws.

TRACK ALONG M-SET LEFT TO RIGHT WITH COLOUR CYCLING

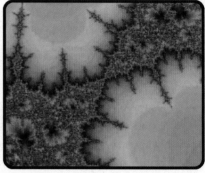

Now, what the Mandelbrot Set and chaos and related things have done for us, is to show that you can have both at the same time. So it's not whether God plays dice that matters, it's how God plays dice.

BENOÎT MANDELBROT

I can tell you exploring this set I certainly never had the feeling of invention, I had never the feeling that my imagination was rich enough to invent all these extraordinary things. I was discovering them. They were there, even though nobody had seen them before. It's marvellous, a very simple formula explains all these very complicated things! So the goal of science is starting with mess to explain by simple formulas. It's the kind of dream of science. And in this case the dream is implemented in a fantastic fashion.

M-SET SECTION WITH COLOURS CYCLING

ARTHUR C. CLARKE

Often when I am looking at my computer screen and watching the beautiful images unfolding, I am reminded of Keats's famous lines: 'Charmed, magic casements, opening on the foam/ Of perilous seas, in faery lands forlorn …' The Mandelbrot Set is indeed one of the most astonishing discoveries in the entire history of mathematics. Who could have dreamed that such an incredibly simple equation could have generated images of literally infinite complexity? We've all read stories about maps that revealed the location of some hidden treasure. Well, in this case, the map is the treasure!

M-SET AND JULIA SET ZOOMS WITH FULL CREDITS AND ACKNOWLEDGEMENTS

END CAPTION OVER M-SET ZOOM OUT:

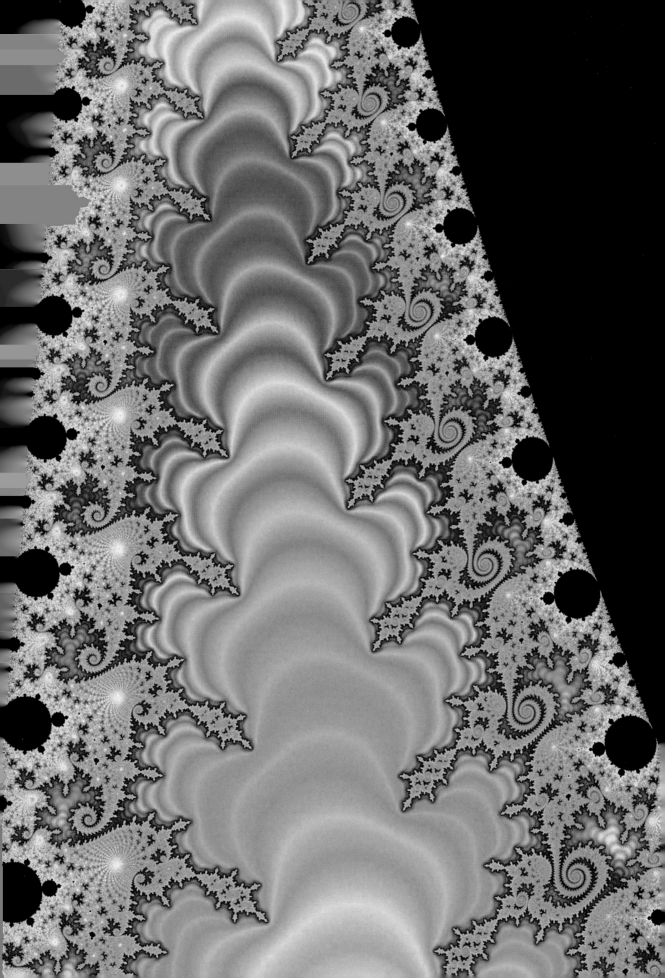

And what you ask was the
beginning of it all? ...
And it is this:
Existence that multiplied itself
for sheer delight of being,
so that it might find itself
innumerably.

Index

N. Lesmoir-Gordon (ed.), *The Colours of Infinity: The Beauty and Power of Fractals*,
DOI 10.1007/978-1-84996-486-9, © Springer-Verlag London Limited 2010